The Story of HOMO LOQUENS

How We Have Mentally Turned into a Species Distinct from Homo Sapiens

The Story of HOMO LOQUENS

How We Have Mentally Turned into a Species Distinct from Homo Sapiens

Dan M. Mrejeru

Published in the United States of America
ISBN 978-1-953904-21-8 (Paperback)

A Terrestrial's Mind Publishing
222 West 6th Street
Suite 400, San Pedro, CA, 90731
www.stellarliterary.com

Ordering Information and Rights Permission:

Quantity sales. Special discounts might be available on quantity purchases by corporations, associations, and others. For details, contact the publisher at the address above.

For Book Rights Adaptation and other Rights Permission. Call us at toll free 1-888-945-8513 or send us email at admin@stellarliteray.com.

CONTENTS

Part Two: Justifying The Hypothesis Of Homo Loquens

Hominins evolution, as toolmakers, was meager when compared to the colossal development generated by the technology cf language with which another but distinct species (Home loquens) built their imaginary civilization.

FOREWORD

Three years ago, in August 2015, I self-published a book titled Solovki Ersatz (the subtitle was "On the Evolution of Modern Human Brain"). When I started to write that book, I was confident that, as the writing goes by, I will be able to answer the main questions generated by that subject.

The writing somehow went fast. I finished the book but in the end, I realized that I failed to provide some good answers. The idea of the book was that some geomagnetic excursions of the Earth's geomagnetic field had influenced human biology; producing some mutations which, in this case, would be responsible for the modern human brain. I still think today that the geophysical argument of that book was, in many respects, accurate or correct—meaning that the increased cosmogenic radiation penetrating to the ground has generated several while very important mutations in the human brain.

As I mentioned in that book, between eighty thousand years ago and two thousand years ago we had nine (nine) geomagnetic excursions where their effective duration covered 30 percent of that time interval. The bulk of those excursions occurred between forty-one thousand years ago and twenty-five thousand years ago when the Earth's geomagnetic field was in excursion mode for almost eight thousand years that was 50% of that time interval.

During the excursional mode, the geomagnetic field intensity lowers by 40 percent, allowing 40 percent more of the cosmogenic radiation to reach the ground.

Also forty-one thousand years ago was a supernova event followed by a significant burst of energy that probably was responsible for ending forty-four species on Earth.

The second idea of that book was that the main changes in human brain occurred during the geomagnetic excursion named Solovki that was placed by the geophysicists between 7,500 and 4,500 years ago.

The third idea of the book was the *"ersatz"* (a German word) that was intended to define, in a more literary manner, an "artificiality." That was about the artificiality of our society.

Having completed the book, I decided to publish it anyway, but in my mind and my soul, I was very angry and distressed about my short-coming and my impossibility to reach my task.

Thus, after publishing the book, I decided to continue my research until I have a valid answer.

During the past three years, I continued to read very many science articles on this subject, and also I chose to search for all related science domains.

Finally in May 2018, a good answer started to take shape in my mind. Hence, I decided to write a paper where to expose the new "trends" (in the Evolution of Modern Human Brain), which I have discovered while making a multidisciplinary correlation between various scientific ideas presented in several scientific papers.

Now, the **new cause for these "trends" proved to be a very distinct subject compared to my original thought: this paramount cause is the development of human language and changes in human brain entropy**. Still changes in brain entropy, in many instances, are generated by high cosmogenic radiation penetrating to the ground. I found in a Ukrainian study of K. Loganovsky (Ukrainian Academy of Science, Dept. of Psychoneurology) published on the 23rd of June in 2009, a discussion on brain effects of the Chernobyl nuclear accident that indicates that the both **strong and loose radiation effects**

are higher in the left hemisphere (i.e., the seat of language and logic and math).

A recent article in the Guardian (UK) documents the fact that life is shriving now in the area affected by the nuclear accident at Chernobyl (Ukraine).

Researcher Shizuyo Sutotu of Shujitsu Women's University examined data from 120,000 survivors of the atomic bomb blasts since 1950. The analysis showed that survivors exposed to between 0.005 Gy and 0.5 Gy of radiation had lower relative mortality than contrast subjects not exposed to atomic radiation. He concluded that low doses of ionizing radiation are beneficial, producing adaptive responses, like stimulating the repair of DNA damage, removing aberrant cells via programmed cell death.

Jerry M. Cuttler, D.Sc., and Myron Pollycove, M.D. in an article titled "Can Cancer Be Treated with Low Doses of Radiation?" say,

> "However, a large amount of research over the past century on the effects of low doses of ionizing radiation on biological organisms has shown beneficial health effects, called hormesis. Moreover, there is considerable evidence that total or half-body low dose irradiation may cure cancer or significantly delay its progression, leading to a reduction in cancer mortality without symptomatic side effects".

Thus, this article by K. Loganovsky is the **first extensive experimental proof that geomagnetic events with low dose cosmogenic radiation act on the left hemisphere.**

There is known that the right hemisphere has a larger volume of white matter (that connects the neurons placed away from one another) and a smaller volume overall, while the left hemisphere has a grayer matter (that is neurons, which are closed from one another) that is packed at higher density.

The higher number of neurons in the left hemisphere favored communication and hence, language.

In my opinion, low doses of natural irradiation (cosmogenic radiation) affected especially the left hemisphere, while helping the rapid development of language and other cognitive and memory skills.

Other papers indicate, like in the cases of the background radiation, that the radiation plays a stimuli role in human biology in general.

* * *

It is well documented that all animals, insects, microbes, bacteria, and plants communicate through "sound." But in humans, this sound communication became the most important technology in use.

The first question would be about a cause that made stringent the need for communication by a complex language as a near-field type of communication. Such a stringent cause would only be in the range of a threat of extinction.

In the same epoch on Earth (80,000 to 20,000 years ago) 177 species of animals, larger than 10 kg, have disappeared. Were humans one of the species on the verge of disappearing? Was the technology-like-language their salvation?

However, the theories about bottlenecks that occurred some 70,000 years ago and again 40,000–30,000 years ago have been discarded because no genetic proof was available. Even then, some events prevented the human population from exceeding worldwide 10,000 inhabitants until the Late Paleolithic population explosion.

But now, the genetic proof would be the mentioned vestibular mechanism in itself. What was the role of such a mechanism? It brought to humans new options to survive and adapt.

However, this mechanism encountered many challenges from the planetary geophysical events and all of them eventually produced small but accumulative mutations responsible for local changes in the human brain-working-entropy modes.

Language explosion was such a cumulative genetic change, consciousness in itself was the result of the same changes manifested here as entropy suppression.

Human brain entropy shows greater entropy than in the brains of other animals, but the scientists consider that the entropy of a primary state which existed before developing of consciousness has been higher than in other animals. In humans, evolved the capacity to locally suppress the entropy, giving rise to an advanced type of consciousness and a language.

As a result of the said suppression, it was possible to develop hand-in-hand the consciousness and the language. This local suppression is the mechanism that created for us the world of order we perceive. Each suppression acts as a filter that selects a certain type of information that has low information content and appears as "orderly" (very low or no variability). The state of increased entropy in the brain increases subjective uncertainty but produces much richer information content.

Forty-two thousand years ago, humans were in danger to experience an increase of brain entropy (cosmogenic radiative energy can increase or decrease the entropy), **which may have caused a high mental puzzlement**, **leading to extinction?** The adaptation was the suppression of the entropy and choosing order.

Not knowing allows one to see first the unknown as magic, but as a later thought, one may start to imagine about the unknown, producing a physical counterpart of it.

As time passes by, this imagination would build up a new physical reality.

Were nouns sufficient to inflict the conceptualization and practice of magic?

Were verbs those vehicles which transited the magic into imagination?

Psychology finds the same process occurring in young children while transiting the magic of the mental state of early childhood into an imaginative state of later childhood.

PART ONE

EMERGENCE OF MODERN BRAIN

OUR WINDOW TO THE WORLD

The oculomotor system produces eye fixations on an object displayed in a visual scene. The process of fixation spends some time on each evaluation because the fixation must describe to the brain the characteristics of the subject of fixation.

The eye jumps from one fixation to the next because the subject seems to move from one position to another; hence, there is not a continuum to be followed, but we deal with separated points which are the subject of each fixation. The space interval between each fixation is called "gap" or "latency" because the brain does not know what exists or happens between each fixation when the eye jumps from one location to another. The brain must cover this gap by making a prediction and the prediction filling up that space assures a continuity between fixations.

The original gap/latency (estimated at 190ms) in the oculomotor system that deals with visual scenes and the motion of objects are filled up with a prediction about the missing part (190ms) of the motion from a scene. Such a prediction is a backward approach because it fills up a space that has been already traveled by the subject. It is a prediction that refers to the past.

In the case of humans, an additional gap is produced by the speaking language that occupies a dynamic virtual space in mind (that varies from 150ms to 500ms). It is filled up with information that is a prediction about virtual motions and dynamic interactions involved in the conversation or in thinking itself that now becomes associated and is structured by the language itself.

Our brain is set up to show the "continuity" of actions and events. Hence, every occurring gap/latency (that is a discontinuity) must be filled up to restore the continuity.

The language-exclusive-contribution (when verbs are involved) amounts for a gap up to 300ms but on average, this figure was determined to be in the range of 200ms during the perceptual and linguistic work in tandem.

The language changed our understanding about time because the grammar introduced the linearity of language while the verbs became used to signal the case of a time when an action or event occurs.

The oculomotor system is built in all mammals by many millions of years of evolution and generates a gap between present and the past, where the present appears as the first (that is the last) to occur action, while the past appears as the secondly occurring action (that is the previously occurred action). Here, the arrow of time runs from the present to the past.

Now, the verbs are the vehicles of that action that takes place in the virtual space of the mind. The verbs are capable of creating meaning for an "intention" that defines a "future" action. Intention implies a general desire or "plan" to accomplish something, while intent is described as an indication about a firm resolve to get it done. The verbs have allowed the brain to deal with future actions by adding that tense that describes the future.

The verbs carrying the intention or the intent have produced intentional programming, changing the language into a programming tool. All this programming is about inserting "control" over the surrounding reality.

Before the creation and introduction of a vehicle for human intention, the time was running from present to the past.

The **verbs change all that, making it possible to indicate an intentional but forward action that is what defines the future**. The rest of linguistic linearity contributed to generating that "asymmetry" that is the "arrow of time."

A study (2015) by Hamidreza Ghader and Christof Monz, both from Informatics Institute, University of Amsterdam titled "What Does Attention in Neural Machine Translation Pay Attention To?" analyzes the role of nouns and verbs in the mechanism of attention that is implied in translation techniques. This study tried to clarify what kind of phenomenon is captured by attention. It considered that "attention can be seen as a reordering model as well as an alignment model.

There were several questions: Is the attention model only capable of modeling alignment? How similar is attention to alignment in different syntactic phenomena?

The result found that attention can model the alignment in some cases, which are the cases of nouns, and in other cases, the attention goes beyond alignment referring to verbs. It concluded that attention follows different patterns depending on the type of word being used. Here is a split opinion about attention having or having not an influence on learning syntactic information.

"The concentrated pattern of attention and the relatively high correlation for nouns show that training the attention with an explicit alignment is useful for generation of nouns."

However, the above is not the case for verbs where it must be captured other relevant information inserted in the context.

This study, even indirectly, points out to the role of various relevant information of the contexts which stimulated the production of verbs.

I have to mention that "attention is defined as the activity of a set of brain networks of alerting, orienting, and executive output. The coordination of the attentional networks implements cognitive control in a context-sensitive fashion. It is the mechanism of so-called selective attention to deal with the limited capacity of information processing via selectivity" (Desmone and Duncan, 1995).

I tried to suggest that changes in human attention behavior have influenced first the use of nouns, and later, a need for higher mobility in expression led to the introduction and use of verbs.

HUMANS AND THEIR LANGUAGE ARE THE PRODUCT OF NEOTENY

S. J. Gould in *Ontogeny and Phylogeny* (1977) explained: "neoteny is retention of formerly juvenile characters by adult descendants produced by somatic development." He argues that "humans are essentially neotenous." Gould claims that in the animal kingdom "neoteny is an escape from specialization."

Marco Mazzeo of University of Calabria (in his paper Cahiers Ferdinand de Saussure-published in Geneve, Switzerland, 2014), analyzing Gould's work shows that:

> the success of Homo sapiens is due to his scarce morphological, cognitive and linguistic specialization. Advocates of hypermorphosis, by contrast, insist on the idea of Homo sapiens as "super ape" (Mc Namara, 1977), as super specialized primate. They are both convenient evolutive strategies, but their difference lies in the modality and rates: hypermorphosis has "strong immediate advantages," while neoteny has "great macro-evolutionary potential whose price is an initial evolutive disadvantage."
>
> Stating that Homo sapiens is a neotenic primate does not imply that he is the most neotenic

> animal. Human neoteny is comparative, and its theoretical strength lies precisely in its comparative nature. Neotenic ontogenesis is useful and not discontinuistic because it contributes to individuate the position of Homo sapiens in the evolutive tangle of primates and hominids.

According to Gould, most of the species are subject to neoteny. Such eventually can be explained by the processes of negentropy, which define all beings. Neoteny is a form of retardation; negentropy, on the other hand, retards the growing entropy in the biological organisms.

Of course, neoteny manifests distinctly from one species to another.

The reason I introduced here the paper of Dr. Marco Mazzeo is about his contribution in connecting the development of human language with the manifestation of neoteny. He tries to demonstrate that language appeared at humans as a result of some particular developments of neoteny in our species.

It is known that when the children are ten to fourteen months old, they are able to say their first word and in general, at the age of one year, they can speak one to three words. In this stage, these children use more gestures than words. Several months later, the situation changes and there are more words than gestures. This example shows that children are not born with a spoken language, but they inherit some forms of gestures. Gestures are symbolic forms and they predate the language that is the highest type of symbolic expression.

I will cite now some ideas from Jean Piaget theory of cognitive development.

The infants zero to two years interact with the world by simple reflexes and few more intentional movements. This stage ends around age 2 when children acquire language.

The stage two to seven years shows a lack of logic and a very egocentric thought (that is called *centration*). They centrate on one property ignoring the rest of them.

Stage seven to eleven years show the beginning of logical thought that also corresponds to the beginning of schooling; here, the education and development of language expose the mechanism of logic that consists of analysis and synthesis.

Stage eleven to sixteen is named the formal operational when beyond logic appear abstract concepts.

In human evolution, the infants start with gestures and only later, they evolved into a spoken language.

But how the neoteny helped humans to develop a language?

As Marco Mazzeo has considered in his paper, the role of neoteny in the appearance of language refers to several aspects:

- "immaturity of the brain (caused by neoteny) is a learning handicap that greatly aids to language acquisition (as a form of communicating meanings);"
- "neoteny produces a lesser environmental tuning, while later it displays wider evolutive variations (again, this initially favors the development of a helping-tool for communicating);"
- "language production and comprehension emerge concurrently with neotenic brain and body development;"
- "the logic of language learning avoids specialization;"
- "inhibition (that occurs in the normal processing of the brain) can represent the cognitive operator that is the evolutive precursor for language;"
- "there is a strong relationship between inhibition and negation; negation is a language fundamental mechanism that is negative-differential;"
- "neotenic logic: to gain less now in exchange for greater future possibilities;"

- "a stimulus is symbolically organized due to the structure of the language."

From all these characteristics exposed by Marco Mazzeo, I will select several that seem to me to have more impact on human development.

In my opinion, the **relationship between inhibition or suppression and negation** appears extremely important because it indicates that while the development of language, by means of communication, substantially contributed to the expansion of the domain of conscious thinking, it also applied a severe selection of meanings by negating; suppressing; or inhibiting some of them.

Thus, language was one of the fundamental tools that shaped our "world of order" by suppressing the elements that appeared to contain "disorder."

By the same means, the language suppressed the entropy while increasing the negentropy.

A lesser environmental tuning inflicted by neoteny produced an environmental handicap that was only diminished by the introduction of language that compensated by increasing the shared capacity of acquiring environmental information. Also, the language contributed to an increase in the control of the environment. Especially after the Out-of-Africa—when the environments became very different and hard to adapt and understand, the language became an absolute must.

Because the language became structured by its symbolic content, it offered an exercise in primitive logic. One can say that symbolic thinking evolved in humans from visual symbols to gesture symbols for communicating, to sound symbols, to graphic symbols, and written symbols.

In the next chapters, one will see that the "graphic" expression in the human mind seems to have roots in the way the brain processes while reflecting the physical configuration of its own structures.

Gyorgy Buzsaki and Brendon O. Watson indicate that the brain rhythms help produce a neural syntax, and help produce, analyze and store cognitive content.

Hans Jenny (1970) demonstrated that when the sound waves interact with water (as diffraction) reveal or generate symmetric patterns with 2D geometry out of the spherical nature (3D) of the sound. The structure created by the sound imprint on water isquasi -3D in nature and can be considered analogs of the vibrational data within the sound.

"Visual thinking," combined with sound perception, is at the origin of early discoveries and experiments in *cymatic*, which led to early knowledge of symbols, geometry, and arithmetic.

"Visual thinking" and acoustic abilities initiated the use of symbols because the sound of vowels pronounced inside the caves with special acoustics will leave on the fine sand or dust of the floor some specific geometries.

Nonlinearity of thought dealt with an evaluation of multiple patterns of the environment. The patterns display structural plasticity that made them vary according to changes in context. Their structure contains information, and information generates function. Each sound variation (as the context) will generate a distinct shape on that sand (content) placed on the cave floor. Hence, context variation would vary the content.

Maybe the most important endowment of nonlinear thinking was the randomly developed capacity to read an unseen function out of context (like the practice of acupuncture and chakras) or to stimulate functions (like meditation and martial arts). These functions were produced by specific vibrations or pulses of energy present in these systems.

The same mental process led it to be defined as the proportionalities existing in nature. But the matter of proportionality was developed further by the advent of linear thinking.

For example, Michael Leyton made the Theory of Shape, and his theory works by capturing the evolution of the shape (circles, squares, cylinders, etc.) from the building blocks (point, line, plane). His theory indicates how the design can be stored as information, and how it can be transferred from one work to another. Leyton explains:

> there is an equivalence between the concept of geometry and the concept of memory storage because a geometric object is defined as a structure from which one can recover the sequence of actions that created its current state. Our brain will have the mental ability to recover the sequence of actions that created such a current state. All images are a sum of geometries.

The statements of Leyton explain the "Vision Thinking" of the Paleolithic man: the analysis of the shape from the environment can manipulate the information content to obtain interpolations and comparisons, extract simple patterns out of complex ones. The analysis can extract historical information directly from every given shape or subject.

The Paleolithic man was aware of the paramount role of the context and its influence on contents.

The development of language changed all that because it tied up things and processes to its inhered "linearity." And then, the content gradually became more important than the context.

Somehow, the neotenic logic of "less is more" (meaning now less, but where that immediate "little" multiplies significantly in the future) paved the path toward understanding what the "future" would be.

The same neotenic concept is embedded in the Theory of Shape of Michael Leyton, where the most simple or primitive geometries can evolve into the most complex ones.

In the meantime, such mentioned processes reflect only the natural mechanism of complexification.

Neoteny is one particular aspect that belongs to processes characteristic for complexification.

Now, returning to both theories, one of the prevailing neoteny in humans, and the other one of superhuman, in my opinion, both are ignoring many important facts:

- neoteny argumentation ignores a wicker environmental tuning that occurred at the time the Homo sapiens left Africa;
- the superhuman ignores the fact that Homo sapiens was weak and on the verge of extinction when he left Africa.

Comparing humans today to the majority of animals, one could figure out that most animals may represent a stage similar to infancy and childhood attained in human development. So, human neoteny, by this criterion, is very little evolved.

But as I said, the neoteny develops unevenly in each species, and there is a large difference in the manifestation of neoteny between species.

A study of the Croatian Institute for Brain Research (Zdravko Pentanjec, Milos Judas, Goran Simic, Mladen Roko Rasin) combined University of Zagreb, Yale University, and University Medical Center, Amsterdam (Harry B. M. Uylings, Pasko Rakic, and Ivica Kostovic) published in June 2011 is titled, "Extraordinary Neoteny of Synaptic Spines in the Human Prefrontal Cortex." The authors indicate that:

> the largest number of supranumerary synapses
> has been recorded in the cerebral cortex of human

> and nonhuman primates. It is generally accepted that synaptic pruning in the cerebral cortex, including prefrontal areas, occurs at puberty and is completed during early adolescence. We confirm that dendric spine density in childhood exceeds adult values by two to threefold and begins to decrease during puberty. This overproduction, including substantial elimination of synaptic spines, continues beyond adolescence and throughout the third decade of life before stabilizing at the adult level. Such extraordinarily long phase of developmental reorganization of cortical neuronal circuitry has implications for understanding the effect of environmental impact in the development of human cognitive and emotional capacities as well as the late onset of human-specific neuropsychiatric disorders.

Recent data on identified neurons obtained from both rhesus monkey and human showed marked regional differences in the number of spines grown and pruned in the basal dendritic tree of layer III pyramidal neurons. However, the number of spines in the adult cortex was related to functional hierarchy, as was the number of spines overproduced, being highest in the prefrontal cortex and lowest in the primary sensory regions. Most studies indicate that the prefrontal cortex undergoes the largest overproduction, and the lowest rate of elimination of all areas, an observation which is explained by the late evolutionary emergence of the prefrontal cortex.

Comparative analysis of mRNA expression in the prefrontal cortex shows that the dramatic changes in the transcriptome profiles in the human brain are delayed relative to nonhuman primates.

A study by Glen N. Saxe, Daniel Calderone, and Leah J. Morales, titled "Brain Entropy and Human Intelligence: A Resting-state fMRI," published by PLOS/ONE in February 2018, indicates:

> **Brain entropy is calculated as the predictability of single-voxel signals over time**. Here, there is a very close relationship between a number of possible states to a system and predictability of a behavior of the components of a system. Predictability relates to the concept of information density described by Shannon.
>
> In a less predictable time-series, with high entropy, each successive data point adds new information. **In a low entropy time-series, the same information is repeated**. Thus, a voxel whose BOLD **signal intensity varies unpredictably over time would be given a high entropy value**, whereas a voxel whose **signal varies little, or in a highly regular pattern, would produce a low entropy value**.
>
> A central function of the brain is to minimize free-energy. Free-energy minimization is the motivation of intelligent systems to avoid surprising states. The brain minimizes such surprising states. The brain uses Bayesian probability to make predictions of understood causal processes to avoid surprises. A brain with a recognition density that can correlate with a high number of possible environmental/causal events would be much less likely to find itself surprised. The free-energy theory of the brain

> would also associate intelligence with high brain entropy. Such a tendency to maximize entropy via active interference would describe a brain poised to make accurate predictions about a large number of possible outcomes. Intelligence represents access to varied neurological states.

Amihud Gilad of University of Haifa (Israel) in a study titled "Neoteny and the Playground of Pure Possibilities," published in February 2015, explains:

> Though some of our genes are certainly involved in our neotenal being and capabilities, our neoteny has made it possible for us to transcend our genetic limitations. It has allowed our brain extraordinary and long-term flexibility. The delay in maturation affords the human brain with continuity in flexibility and plasticity and allows us to keep on learning throughout our lives. It is not an accident that unusually creative brains often are associated with playful, even- curious immature and playful personalities.

As Leonard Hayflick (PLOS Genetics, 2007) states:

> There is a huge body of knowledge supporting the belief that **age changes are characterized by increase entropy**, which results in the random loss of molecular fidelity and accumulates to slowly overwhelm maintenance system.

Hence, human maturation stage sees increases in entropy. Because of this particular stage development, **overall entropy in humans is higher than in most other animals or primates**.

However, during the Out-of-Africa migration, the human balance because of statistical limitation in aging, generated an epoch of overall low entropy. Much more recently, this balance changed toward higher entropy. Hence, higher entropy in humans is a recent phenomenon.

* * *

There is the case that a weak Homo sapiens that left Africa sometimes eighty thousand to fifty thousand years ago, but these immigrants have with them the developing skills of language.

However, the language evolved in parallel with planning and control, and it also stimulated a higher level of socialization because of the evolving communication capacity.

The success of this migration depended exclusively on the compensation effect offered by the mentioned skills. Also, the neoteny behavior of less-is-more was a vague promising that was intended to work at a later time.

As we know, the migration was a total success, validating the thought that the still relatively primitive language functioned well as new way of communication, and together with planning/ control and higher socialization skills provided the very much needed compensatory element.

As one will see in the next chapters, I was able to define the emergence of the new brain and the explosion of the language skills around forty thousand to thirty thousand years ago.

Hence, it took to Homo sapiens almost twenty thousand to forty thousand years (depending on when out-of-Africa occurred) of migration into Eurasia and Oceania (but only later in the Americas) to produce really a new mental capability.

In sum, the homo sapiens was far from being a superman, and on the contrary, he was very weak when the immigration started. But against all the odds, he prevailed.

INTRODUCTION

What made sound communication in humans so disruptive compared to all other life forms that communicate by sound too?

In my research of the last three years, I discovered language is not as disruptive as a mentioned mechanism developed in the human brain. Hence, the language is secondary to the mentioned mechanism. Or, the **language is the communication carrier of the results produced by the "mechanism."**

In the next eighty pages or so, I will present the correlation I was able to make.

For a visual scene, the brain makes a backward prediction that is the prediction about filling up a gap that is about an eventual past action. Is there any prediction about the future?

It is assumed that most animals make predictions on something that is about to happen. But here we are not certain if such a prediction is only based on a comparison with the past or something else. This case is maybe based on something that we call natural symmetry and is present all around us. Anyone may assume an event simply based on symmetry. In the meantime, we are not certain about the sense of symmetry that can simply be another aspect that the processing of the oculomotor system generates in our mind. We have previously seen that the oculomotor system selects one-in-a-million environmental information. Maybe this system selects only those things which are compatible with its processing, and here, the symmetry can be one of them.

Psychological studies on contemporary primitive people indicate that those people have a very little conception about the future; they mostly live in the present and the past. Primitive cultures are strongly tradition-bound, intensely conservative, resisting to adopt a new practice, while not tolerant for novelties and differences in opinion. All these behaviors indicate how powerful is their dependency on an oculomotor system that evolved in many millions of years to particular attributes.

However, the human language first evolved to nouns and adjectives, producing a virtual static place in our mind. At some point in time, here estimated as forty-three thousand to thirty thousand years ago, it was added to the virtually of the static language the verbs and adverbs which introduced "action" and generated a dynamic virtual space in mind. The verbs made the language "flow," made it complex and made the words turn into a type of instructions, which collapsed high-dimensional mental space into low-dimensional that is the linear but virtual space of language, and which forced the thinking to adopt the same structure—we think now in the format of the language. Even then, the language seemed to have copied a type of genetic structure.

* * *

All decisions are made unconscious; thus there must exist an "a priori" selection that is unconscious and remains unconscious. Only the unconscious result that generates the "action" becomes conscious. We are aware of our actions.

The unconscious as the conscious does shows stratification of various levels of selected information where all levels are displayed hierarchically. It could be suggested that this natural type of organization, when it is analyzed from the base to the top, indicates an increasing amount of variation. The top of the hierarchy shows the most variation or a fast-changing behavior.

This top point is on the edge of chaos that is the place where disorder and high variance switch downward into the increasing orderly formats. In short, this place of transition is the locality where entropy becomes diminished in an extreme or very sharp manner. Several measurements found that the local entropy can be diminished in the range of one million times or even more when compared with the surrounding entropy.

As appears, the unconscious differs from conscious exactly by this difference generated in the selection of variance. The unconscious shows a selection of higher variance that stays in contrast to the conscious that is all about driving out or selecting out the "variance" along with its "unpredictable" behavior while driving in "order" and "predictability."

Henrique de Oliviera (October 1, 2014) in his paper, *Axiomatic Foundations for Entropic Costs of Attention* explains:

> The allocation of attention can be understood as an information acquisition problem. The decision-maker faces a menu of options, each yielding a consequence in each state of the world. When choosing what information to pay attention to, the resulting posterior is not known, yet; only the distribution of possible posteriors is known. It results in the reduction in uncertainty from that intended observation, where its Shannon Entropy gives the uncertainty of a probability distribution.

Here, I must clarify the correlation between information and entropy. The mathematicians would demonstrate to you that the level of uncertainty would determine the level of entropy. Maximum of uncertainty has the maximum of entropy, and also it has the maximum of information. The bad news is that zero uncertainty is zero information and zero entropy.

In the abstract of this paper, I mentioned that our oculomotor system receives eleven million bits of information and selects out of this total only forty bits or so. Now, we can connect this reference to information and entropy: if the mentioned process of the selection made such selection, it reduces the information amount and entropy by a million times.

However, all research indicates that creativity refers to higher levels of embedded information, an associated higher level of entropy where there is a significantly higher than normal amount of variation and uncertainty; it might turn obvious that this higher information content is located in the hierarchic distributions of the unconscious level.

It should be mentioned that natural hierarchies appear quite distinct compared with our structures that we design for building hierarchies. Human managers categorize all components into clearly defined categories so the workers (agents) can easily and rapidly distinguish and use components; organizational information is highly structured as well and clusters around job roles and skills. These structures affect the efficiency favorably, but they do a routine.

The research on creativity found that if the components are not already pre-organized into distinct categories, then the information can exist as a flat structure, increasing cognitive flexibility and allowing to combine information from distal conceptual categories. On the converse, the hierarchic organization obstructs all these processes seen as creative because all processes are following the predetermined while embedded structure. There is no place for flexible moves.

One might conclude that the selection of at least the unconscious information in our brain is organized in a dual mode where simultaneously should exist stability combined with flexibility. It might be a dynamic process where the two states of matter intertwine. Creativity is defined by the very moment when one can experience a sense of flow that appears out of nowhere but against the normal process that is trying hard to form an idea and very poorly succeeds with it. Creativity is that flexibility that deals with quite a multiplicity of patterns or ideas which are spread on the entire neuronal network.

In everyone's mind occurs an "action-perception cycle" that repeats itself every 300–400ms. This cycle seems to indicate a particular structure that has been developed in our brains, reflecting the aforementioned dual structure. It might reflect the rate of change from one type of entropy to another that repeats itself almost three times per minute.

Freud in his research has found the presence of the same dual character showing that unconscious appears as a mixture of timelessness and time and,in general, as a mixture of all no-contraries vs. all contraries and polarization. One can say that here sometimes is a pattern (that is "order"), but other times there is no pattern at all (that is chaos or disorder). In short, we deal here with various levels of entropy whereas the entropy rises, the disorder increases. This is the place where the information navigates from one higher level of entropy toward one lower level. At higher entropy, there is timelessness; at lower entropy, the time appears. At higher entropy, there is no repetition, no proportionality, and no regularity because everything is disordered. Entering the lower entropy environment, all attributes of "order" come into place: repetition, proportionality, symmetry, regularity and predictability.

* * *

The unconscious screens shape objects that "you" (where "you" represents your own conscious) want to "see." So, the unconscious must know what the goal is or what is relevant for the conscious state, and it fits that demand efficiently and effectively. It is done by filling the language-related latency with "unconscious goals," which must be connected with the subject on trial.

The information of the dynamic virtual space of language has an abstract content associated with the virtual thoughts and interactions, as being expressed in conversation and thinking. This same virtual space became the host of all abstract thinking and art. The abstract pictures the uncertainty and is rich in information. There is to assume that unconscious generated all "abstractions," which have been passed to conscious expressions.

Language has been generated by the developing conscious mind and influenced thinking by forcing it to follow the rules adopted by the language.

It is like, one may need something but one does not know how to make that something available. And then, the unconscious fulfills that goal with needed information because the unconscious has access to that information with higher uncertainty that has more information content and has higher entropy.

The research on illnesses with attention deficit indicated that attention shows a complex mechanism that covers all rhythms present in the brain. Paying attention may increase some of the rhythms frequency and simultaneously it may decrease other rhythms. When this discrepancy increases between the connectivity of various neuronal networks then, the illness arises.

However, in human evolution through language development, the role of attention increased too. Increased attention reflects an increased type of local entropy of a particular neural network. The increased attention brought in more information, helping humans to "discover" and concretize some previously unknown information. The discovering process was shaping up new ideas.

* * *

This dynamic virtual space of the mind made humans unique and it was entirely due to the verbs added to the language.

Before the introduction of verbs, it was in mind a virtual space filled with nouns and adjectives while the sense of action was still provided by gesturing.

As previously mentioned, the oculomotor system generates language-related latencies caused by the jumps manifested between the eye fixations that follow the virtual motions developed by conversation. The mind must produce a continuum of that conversation, and it fills the mentioned gaps with unconscious information.

However, the dynamic space of the mind where the conversation takes place can maneuver the elements implied by that conversation in a linear approach and can also find a place for the dynamics of the intent. The intent becomes the future, and a particular tense describing the future defines it.

* * *

In the next chapter, it will be a more detailed description of the brain functioning and some indications will be provided about the brain mechanisms which helped develop the human language and transformed it into our best technology.

Psychological scientist, Juanma de la Fuente of the University of Granadain Spain, did several experiments about how we perceived time and discovered that it depends on whether we are focused on the past or the future. His experiments suggested that for most young adult Spaniards, the future is lying in front, while for the elderly Spaniards, the past is associated with the front. In another experiment, almost half of Spanish college students perceived the past as being in front. Also, other experiments proved that the Arabic speakers in Morocco tend to see the past as being in front of them and the future behind of them.

Now, taking the assumptions of the quantum mechanics—it would mean that a particle could carry the action of its measurement back in time to when it was entangled, affecting its partner. Here, the causality runs backward and this is a similar process with that occurs in the oculomotor mechanism.

The newly created (probably thirty thousand years ago) dynamic virtual space of language was adapted to the existing oculomotor mechanism but plays a secondary role here, while the **changes applied to the oculomotor mechanism are de facto technology**. But they work together in an almost undistinguishable association.

Hence, unconscious appears to operate its selection mechanism that also controls the conscious selection. It is a goal-oriented-processing that is called "*unconscious bias.*"

From that point on, humans became distinct from all other animals. As many scientists said, **it was not a natural process**. We will see about that because numerous genetic mutations occurred in the last forty-two thousand years, and they seem quite influential.

One can consult the book, *The User Illusion,* by Tor Norretranders from Denmark that, as Hazel Gale said in *Seeing Things* (January 2015), "...asserts that 90% of what we see, hear, feel, etc. is internally generated because the vast majority of what we think we perceive around us is actually fabricated and added by the brain before it's consciously recognized."

John Cast commented, "Consciousness represents only an infinitesimal fraction of our ability to process information. Although we are unaware of it, our brains sift through and discard billions of pieces of data to allow us to understand the world around us. What our consciousness rejects constitutes some very valuable information." It is the information we had very little recovered but enough to produce civilization. No other animal has the mechanism to do this.

* * *

For Manfred Steger and Paul James assert, "imaginaries are patterned convocations of the social whole. These deep-seated modes of understanding provide largely pre-reflexive parameters within which people imagine their social existence-expressed, for example in conceptions of the global, the national, the moral order of our time. **The illusion of coherence, control, and totality** is by no means unnecessary or inconsequential (as something that is illusory). Imaginary identifications can have very real effects."

At the apex of this "*imaginary strategy*" introduced in human thinking some 3,000–2,500 years ago, the multiplication of imaginary options increased from fast to exponential, producing what we call "civilization."

In a paper published on October 27, 2017, entitled, *What is in the Name-Asia* the authors Edward Dawson and Peter Kessler suggest that the Indo-European tribes (spreading from Yamna Culture and calling themselves Aryaman that is a term that was defining "the man of the truth") from Pontic and Caspian plains adopted "a new type of philosophy" (3500–2500 BCE) of the daily life and spiritual life devoted to the truth as it is.

In my opinion, this new cultural trend signifies the adoption of "materiality" as it was. While it appears, the materiality evolved significantly between 5000 BCE and 3000 BCE, also creating an egalitarian society in gender. Some historians deny the existence of a full matriarchate, considering that this was a strictly local issue with more or less of one single gender dominance.

But this "materiality" already progressed into metallurgy and maneuvering of chemical substances, both implying processes of analysis and synthesis. Nevertheless, agriculture was quite evolved at that time. Hence, it was the recognition of an existing "materiality" that was based on "logic thinking," while the logic itself was the result of "imaginative thinking."

According to historian Marija Gimbutas, this was the moment when the war-like Yamna Culture destroyed the neighboring egalitarian matriarchal cultures like Cucuteni-Trypillian Culture (around 3000 BCE), which, at the time, it was already evolved into significant material development. The war-like Aryan cultures of the Pontic-Caucasian Steppe that was spreading (from 3000 BCE to 2000 BCE) to Caucasus, Anatolia, Central Asia, and South Siberia before to proceed to the conquest of India and the Middle East introduced the patriarchate that allowed the dominance of men.

In the meantime, this new trend indicates a turn from traditional but ancestral "emotional thinking," that created "a culture based on direct experience," toward the incipient logic.

However, the Athenian philosophers were able, in the fifth and fourth centuries BCE to generate a theory of logic thinking.

The literature defines "logic thinking" as an "imaginative type of thinking." In the 2,500 years that followed, the "imaginative thinking" gradually evolved in the "linear progressive thinking." Since the "Industrial Revolution," when the basis for manufacturing of "mechanisms" was created, the "progressive thinking" entered an exponential evolution that accelerates as the time pass by.

Today, one can say that the conclusion of human evolution into civilization refers to the creation of the "**digital universe**" that in 2020 will reach 44 zettabytes (ZB) that is 44 trillion GB of data and which means 5,200 GB of data for every inhabitant of Earth. Its size is doubled every two years, and it is growing 40 percent per year for the next decade.

But unlike the physical universe, the digital universe is created and defined by software that is a human-made construct.

In 2013, only 5 percent of data created had been analyzed. In 2013, less than 20 percent of the data in the "digital universe" is either stored, even temporarily. By 2020, that percentage will double to 40 percent. Most of the "digital universe" is unsaved and mostly discarded. In 2013, the capacity of storage could hold 33% of the "digital universe," and in 2020 that capacity will diminish to less than 15 percent.

Almost all data generated is machine-generated data that is machine-analyzed data, and this proportion exponentially increases in every single year because the number of machines which do the analyzing also increase exponentially.

The investment in spending on IT hardware, software, services, telecommunication, and the staff is an investment in the infrastructure of the "digital universe," and it grows much faster (5% and faster) than the rest of the economy in the developed countries (where annual growth is 2–3 percent).

Entropy means a process where "order" deteriorates with time; an exponential increase in data production rends huge useless swaths of recently acquired data which cannot be stored and analyzed.

In the meantime, digital data tends to be highly controlled because this is the reason for its production. It is a "digital universe" that provides information about our world of "order." This fundamental human-made design makes the "digital world" a low entropy world. We collect such data to eliminate its inherent "uncertainties" to prevent its natural variability to manifest; otherwise, our machines become "unpredictable" and useless.

Our world has a locally manifested type of one-in-a-million-times less entropy than the natural world that surrounds us. Our machines must run at the same low-level entropy for us to understand their work and their results. These mentioned machines generate our "digital universe."

THE LOOKING GLASS OF LANGUAGE

The evolution of human language seems to indicate a lot of information about the changes that occurred in the human brain since the language revolution placed some forty thousand to thirty thousand years ago.

I will start with several quotes from the study "Do Low Doses of Ionizing Radiation Affect the Human Brain" published on 23 June 2009 by K. Loganovski with the Institute of Clinical Radiology of Academy of Medical Sciences of Ukraine.

The study analyzed the results of radiation on the brain of the people in the Chernobyl area who were subjects to different levels of radiation caused by the nuclear accident.

The study compares the results from strong radiation and low radiation. In strong radiation cases, the clinical results indicate most damages to cortex area of the brain and left hemisphere. This fact shows evidence of left-brain sensibility to radiation high and low. This is the first medical argument I found about the possibility that the cosmogenic radiation stimulated stronger the left-brain, generating the language and all other associated features like logic, symbols, math, etc.

However, here come now the studies about the subjects living in high radiation background areas. These studies found the background radiation to stimulate the immune and other systems in the human body. No research on background radiation effects on the human brain have been published, yet. Making a correlation with the other body systems, one can admit that the brain should present similar effects which are in the category of stimulants.

Here I recommend for whoever is interested in this subject to read another paper I wrote and I uploaded on academia.edu under the title "Augmentation and the Illnesses of Civilization."

* * *

We know that up to a certain point, the language was formed only by nouns with later addition of adjectives. We also know from the study of primitive languages that they evolved rapidly to large complexities, and the grammar rules were added.

We know that prehistoric or Paleolithic humans have a brain developed for visual thinking that was a brain implying nonlinear processing. It was probably a type of parallel processing.

All old and proto-languages show a high level of complexity. Let's see what may have caused that complexity. The linguistic research agreed that all languages are linear.

So, at that beginning of the language development, it occurred a combination between visual/nonlinear thinking and the linear stringencies of the language. We can say that nonlinear is based on the absence of rules or regulations. Hence, the mentioned combination of the two systems has rapidly produced linear additions, which were very little regulated and generated an irregular aspect of the language.

Whatever has been added, it was on a single line crowding that line with linguistic information. It made the language become very complex.

Here, we have to consider that the approach of parallel thinking was gradually disappearing from human thinking, but this process took a very long time. I would say that the linear rules of grammar gradually gained more power, and this process of regularization was producing a simplification or reduction applied to previous linguistic complexities. In a paper of Ray

Jackendorff and Eva Wittenberg titled *Linear Grammar as a Possible Stepping Stone in the Evolution of Language* published in *Psychonomic Bulletin& Review* in February 2017, the authors explained:

> We suggest that one way to approach the evolution of language is through reverse engineering: asking what components of language faculty could have been useful in the absence of the full complement of components. We explore the possibility offered by linear grammar, a form of language that lacks syntax and morphology altogether, and that structures its utterances through a direct mapping between semantics and phonology.
>
> We find evidence of linear grammar in a wide range of linguistic phenomena: pidgins, stages of late second language acquisition, home signs, village sign languages, language comprehension (even in fully semantic languages), aphasia, and specific language impairments. We also find a full-blown language, Riau Indonesian whose grammar is arguably close to a pure linear grammar. In addition, when subjects are asked to convey information through nonlinguistic gestures make use of semantically based principles of linear ordering. Finally, some pockets of English grammar, notably compounds, can be characterized regarding linear grammar. We conclude that linear grammar is a plausible evolutionary precursor of modern fully syntactic grammar, one that is still active in the human mind.

However, in languages, the words and grammars change and mutate over time, and new versions rise to dominance while others face extinction. In each language, different but many rules competed over time. The irregulars face massive pressure to regularize and conform to new norms.

Now, all languages tend toward regularization that makes them increasingly subdue to new verbal/linear thinking.

The grammars imply synthesis, which is a process directed linearly. Here, the linguistic synthesis paved the road toward more synthesis in thinking that has led to logic approaches. The language was the tool that made humans break the wholes into parts.

I have to mention that linguistic structure and its linear rules gradually came to alter the original visual/nonlinear thinking; changing the structure of thinking into a verbal, synthetic, and linear dependent type of thinking. We call it today the verbal/linear thinking.

An import element in the evolution of languages was the time concept. As in visual/nonlinear thinking, the time was intimately related to space description. Hence, the languages developed concerning the environment. Here, a location can influence the context of an utterance.

As the language evolved, the three-dimensional spatial approach (left, right, high, low) was gradually replaced by a horizontal linear approach (front and rear).

It must be said that even when not all languages have verbs, all languages find many ways to express action and tenses without relying directly on verbs because nouns and verbs both have past and future tenses. However, when nouns are embedded in taxonomy; this non-verb structure will not allow performing any action, except that one embedded in taxonomy. A semantic (meaning related) hierarchy is a taxonomy. The action was provided by the semantic (meanings) of things, which share the same characteristics, and the action was limited to such meanings: like repeat an action to locate an element of similarity, or extending a pattern into a more general structure.

This fact shows that our Paleolithic ancestors had a limited array of actions before the introduction of verbs. The verbs produced input and output.

The nouns and verbs are processed in distinct areas of the brain, but when the same word can have two meanings, like one of the nouns (example: walk), and one as a verb, they are processed in the same area of the brain. The verbs become activated in neural areas associated with the movement. There is some relationship between grammatical class and semantic representations.

Now, I like to cite M. Bendy, A. Caramazza, A. Pascual-Leone, and R. Saxe, all with MIT and Harvard Medical School and their paper, *Typical Neural Representations of Action Verbs Develop without Vision,* that was published by Oxford University Press (2011) and Oxford Journals (February 2012). I will quote several ideas expressed there and important for my paper.

> Many empiricist theories hold that concepts are composed of sensory-motor primitives. For example, the meaning of the word "run" is in part a visual image of running. If action concepts are partially visual, then the concepts of congenially blind individuals should be altered in that they lack these visual features.
>
> We find that the representations activated during word comprehension are not altered by congenital blindness. The preserved neural substrates of action verb comprehension stand in contrast to the striking plasticity in sensory brain regions following early changes in sensory experience. For example, the location and functional profile

of visual brain regions are changed by congenital blindness from vision to audition and touch and even to higher cognitive domains. Our data suggest that these sensory changes do not carry forward into conceptual systems.

It is nevertheless still possible that there are privileged relations during development, between conceptual domain and specific sensory-motor systems. There is the possibility that the perceptual systems are partly organized along conceptual domains. There is also possible that the conceptual brain regions do not exhibit experience-dependent plasticity.

We, therefore, favor the interpretation that both conceptual and perceptual brain regions could exhibit experience-dependent plasticity, but the development of conceptual brain regions is robust to the absence of vision in particular. That is, the experience of blind children is not different from that of sighted children in ways that matter for the formation of brain regions involved in understanding action verbs.

The present neural data suggests that at least in the case of action verbs, not only the content but also the format of semantic representations is similar in sighted and blind adults, and may be influenced by linguistic experience.

Our data suggest that concepts retrieved during action verb comprehension areabstracted away from sensory-motor experiences and represented in a modality-independent format.

The reason I brought several extracts from the paper above is that it allows several independent conclusions on my part.

One idea is about the *action concepts,* which are found here like being *not related to the visual images*.

The second idea is that *the neural substrates of action verbs do not suffer changes because of changes in sensory (visual here) experience* like because of congenital blindness.

The third idea is that the *perceptual systems are partly organized along conceptual domains*.

In my opinion, the mentioned study indicates that nouns and verbs are produced by several genetic mutations that changed the way the human brain worked.

Hence, the congenital blind individuals will continue to benefit from these genetic changes, even when they lost the sight.

The same study also analyzed the animal nouns (being considered here as having the highest potential motion among nouns) against the action verbs (high-motion verbs) and the mental verbs (low-motion verbs). When all categories (like sighted and congenital; blind; children and adult groups) have been compared side by side, no significant differences have been found.

In the meantime, all perceptions seem to work in tandem as being genetically interrelated, so the loss of one perception at birth is compensated by the activity of other perceptions. This possibility suggests how the mentioned mutations became interrelated to conceptual domains and cognition.

This genetic opinion of mine would support Noam Chomsky's theory on the innate capacity for language in humans. Therefore, the language evolved along several brain mutations, and it generated an interdependency between all these mental features. It may also show that based on the same genetic background the language can evolve on its terms or somehow independently of perceptions. This approach may support another concept that is that of artificial intelligence and computer languages.

In short, the congenital blind people's brains (as tested by fMRI) performed in all categories similarly to normally sighted individuals because of an existing identical genetic matrix.

However, the introduction of action verbs allowed a process of programming because it allowed an input followed by output. Since the grammar of language imposed linear rules, it changed the thinking itself that became dominated by verbal rules.

Junichi Toyota in a paper titled *On Binary Feature in the Evolution of Human Language* published in vol. IX, 2012, of Ed. Nicolaus Copernicus University said:

So, it can be claimed that the verb originated from the active noun, and the inactive noun was simply retained. Note that active nouns stayed as nouns, too, but the action implied by them forced the verb to emerge.

Once we have a binary opposition of noun and verb, the base for complex grammar is made. The category adjective emerged as an intermediate category between noun and verb.

Here, it is important to define the presence of two types of verbs: mental and action verbs.

Mental verbs act somewhat differently than active verbs:

- mental verbs have meaning that is related to concepts, like discovering, understanding, thinking, planning, and so on;
- action verbs express physical or mental action;
- most mental verbs are not used in progressive tenses;
- when mental verbs are used in the progressive, their meanings may be different from when they are used in the simple tenses;
- mental verbs are used with the present perfect rather than present perfect progressive (such as an action that started in the past and continues to present).

I like to say that *mental synthesis is a conscious, purposeful process of synthesizing a novel mental image from parts stored in the memory*, and is different from simple memory recall and dreaming.

In the study titled *Cognitive Revolution,* the author explains:

> Evolution of figurative arts, creativity and innovation, design and construction, religious beliefs, and steadfast colonization of the planet are highly correlated with each other in time and geography and are indisputably associated with Homo sapiens diffusion out of Africa around 70,000 years ago. This abrupt change toward behavioral modernity has been characterized by paleoanthropologists as the "Upper Paleolithic Revolution," "Cognitive Revolution" and the "Great Leap Forward" and is unlikely to be caused by the acquisition of modern speech apparatus more than a half million years earlier. Rather, the sudden change in human behavior is consistent with a single enabling process causing these multifaceted behavioral changes. *This enabling process may have been the acquisition of lateral prefrontal cortex as a driven constructive imagination, called mental synthesis.*

In my paper, I call such a sudden enabling process as a mutation in the vestibular mechanism named oculomotor system, and it mostly resulted from an introduction of a sustained and motivated type action in our prehistoric doing around thirty thousand years ago.

The language with verbs allowed the programming of rules into other people minds. As a result, in late Paleolithic, the human societies became governed by very strict and conservative rules. It was a combination between irregularity (as no control) and regularity (as of control). The society mimicked the complexity and parallel approaches achieved by language in the same period that was combined with the irregularity of the rules (competing rules), which sometimes nullify one another. They wanted to control, so they made a rule; but the rule, let's say, proved no control effect. Hence, they made another rule with the hope to achieve better control. It was the trial-and-error period. And some of the rules had a parallel disposition because the mind still used the inherited nonlinear system. For our current linear thinking, such parallelism appears irregular.

The descending into Mesolithic-Neolithic epoch started to bring changes in language evolution which were directly reflected in social organization. Some of the new rules have proved to be more effective than the previous ones, and gradually the simplifications (discarding of some ineffective rules) came on the scene in language and society. Also, the parallelism of rules became gradually discarded.

The evolution in language and society were both directed toward better control of everything (of speaking itself, of people's minds, and of surrounding environments), and the action verbs were paramount in achieving this ultimate scope.

Today science can use a new tool in analyzing the *language that is the semantic entropy*. It analyses a word by interpreting its entropy that represents its ambiguity and is inverse proportional to its information content. *The most useful words have the highest entropy. In the west, the nouns have the highest entropy. In the east, the verbs have the highest entropy.*

This mentioned measurement produces an entropy ranking that correlates well with language dynamics and evolution.

Studies on semantic entropy of children have shown that children until the age of three have a high ratio entropy between noun-verb like it was theorized for our prehistoric ancestors.

The Westerners still preserve as adults this high ratio noun-verb, while the Easterners have as adults a low ratio noun-verb, meaning that they used much more the verbs than their Westerners counterparts.

LIFE WITHOUT LANGUAGE

It will be interesting to find out what today's specialists think about the time when humanity did not use a language.

On this subject, which bears certain significance for defining human thought as it was before the advent of language, I found in Neuroanthropology magazine an article of Greg Downey published on July 21, 2010.

The author said:

> Thought without symbols-life without language-it's a cognitive reality that is virtually impossible for most modern humans to fathom. For the vast majority of us, out thought processes have been profoundly shaped by the introjection of language into our cognitive worlds." "What would human thought be like without language?
>
> The question of the relationship between language and 'mind,' or between symbolic resources and cognitive abilities, is hard to address in anything other than the hypothetical.
>
> In fact, most historically recorded cases of feral children, however, suggest that they do not develop any language ability at all, perhaps even failing to develop symbolic abilities." "The rare case of individuals without language offers some

> potential window in on life across the intellectual Rubicon if we had developed mentally without immersing ourselves in the shared symbols and the communicative reality of language.

The author quotes intensively the writing of the researcher Schaller who studied a subject without language. The author continues: that many languages do not treat time as an abstract, spatialized, undifferentiated flow but highlight differentiation, seasonality, and sequence. Anthropologists still tend to agree that understanding of time can differ and that Western treatment of time as a kind of flow through undifferentiated, measurable durations is just one version of inflection of the sense of time with its own distinctive emphases. Time, for example, may be difficult to perceive in certain ways if you are not culturally trained to habitually conducting yourself about time appropriately: certainly, there is a deep cultural difference in the degree to which people orient themselves by the clock.

Language diversity itself is an insight into human cognitive diversity. Every language is an outcome of how the human brain works. Each language has used the brain in a slightly different way.

Language was not simply an addition to the cognitive repertoires; it may have displaced or disrupted other forms of thoughts and interaction.

We tend to think of ourselves as cognitively complete, profoundly abled, without limit. But clearly many languageless individuals had to find some way to compensate for their deficits, whether it was through mimetic thinking or through some other constellations of adaptations. This languageless cognition would not be simply prelinguistic, childlike thought because adult languageless individuals function much more adeptly than four-years-olds. But how this nonlinguistic, adult cognition might operate, what it might include?

Some things appear to be easier to "get" without language, other things appear to be a kind of all-at-once intuition.

I have to add to the above quotes some explications.

The language developed further the consciousness. Language made people think-the-way-they-talk, changing the structure of human thinking completely.

In a person's life, the language evolves together with cognition from infancy and childhood, demonstrating a progression through developing skills.

Maybe the most interesting aspect being raised by Greg Downey's article is about this relationship between language and consciousness.

Most specialists would say that a person that is languageless would have a "primitive consciousness" probably similar to our ancestors from Paleolithic. The thinking of the languageless individual would be very different compared to an individual who possesses language. The language improves the cognition and memory. Hence, a languageless individual will not have those abilities.

Another essential distinction would be the fact that our Paleolithic ancestors, until the explosion of language some thirty thousand to twenty thousand years ago, had a larger brain because their bodies were providing such an increased level of energy. Now, the higher energy is considered to be involved in dealing better with "uncertainty."

So, the Paleolithic individuals by their stage of evolution would have a very distinct way of thinking that was based entirely on visual/nonlinear thinking.

Our modern brain shows the conversion produced by the implementation of the language that drove us into linearity.

The last issue in the languageless individuals was about the perception of time. As described by the author of the article, they do not perceive time in a continuous flow as we do (but mostly due to education and culture biases).

I was curious myself to find out how a languageless individual sees the issue of 'future,' but the article did not mention anything about it.

THE BRAIN INSIDE OUT

Recently (July 30, 2018) an article by Jennifer Ouellette under the title, *A Math Theory for Why People Hallucinate,* was published in Neuroscience. Her article is a correlation between several novel types of research on this matter of Brain Science.

I thought that this article serves very well the subject I will present in my paper. Thus, I took the latitude to quote several very important ideas.

> In 1979 Jack Cowan of the University of Chicago set out to reproduce those hallucinatory form constants mathematically, in the belief that they could provide clues to the brain circuitry. Cowan and his graduate student Bard Ermentrout reported that the **electrical activity of neurons in the first layer of the visual cortex could be directly translated into the geometric shapes people typically see when under the influence of psychedelics**. The math of the way the cortex is wired, it produces only these kinds of patterns," Cowan explained recently. In that sense, *what we see when we hallucinate reflects the architecture of the brain neural network.*

Here, I like to add that for some time it has been assumed that some of the rock-painting of Australia was the result of turning inside-out of the brain of primitive aboriginal artists. It also was assumed the involvement of hallucinogens.

The Red Lady of La Miron (Spain) is a skeleton dated nineteen thousand years ago which shows hallucinogenic mushroom in its teeth. It is the first evidence of Paleolithic use of hallucinogens.

> A collaborator of Cowan's, the physicist Nigel Goldenfeld of University of Illinois, Urbana-Champaign, has proposed a twist on the original idea that factors in noise. Experimental evidence reported in two recent papers has bolstered the theory that this stochastic Turing mechanism is behind the geometric forms constants people see when they hallucinate.
>
> Images we "see" are essentially the patterns of excited neurons in the visual cortex. Light reflecting off the objects in our field of view travel to the brain, as signals, and stimulate neurons in the visual cortex in patterns that normally mimic the patterns of light reflecting off objects in your field of view. But sometimes patterns can arise spontaneously from the random firing of neurons in the cortex-internal background noise, as opposed to external stimuli.

Peter Thomas, an ex-collaborator of Cowan, explained:

> the image undergoes a transformation of coordinates as it is mapped onto the cortex. If a neuronal activity takes the form of alternating stripes of firing and non-firing neurons, you perceive different things depending on the strips' orientation. You see concentric rings if

> the strips are oriented one way. You see rays or funnel shapes emanating from central point-the proverbial light at the end of the tunnel common in near-death experiences. And you see spiral patterns if the stripes have a diagonal orientation.

In my opinion, the activity of some genetic structures in the brain suffers a neuronal excitement (electric), which could produce associative sound effects, and which could be associated to particular meanings developing in mind. The uttering of such associated sounds has been at the origin of nouns in our language.

A condensed matter physicist by training, Goldenfeld and his graduate student Tom Butler in 2008, showed in their research that when a herd's population is relatively low, random fluctuations can have big effects, even leading to extinction.

Here is an eventuality that could describe the evolution of extinction crisis and which crisis also developed in the hominin species, making all other hominins to disappear, excepting humans. It is estimated that even humans have been many times very close to extinction. Taking into consideration the length of the time scales on which these phenomena evolve, no one could presume that humans are not still due to extinction.

Goldenfeld showed that "to function properly, the visual cortex must be primarily driven by external stimuli, not by its own internal noisy fluctuations."

Other studies show that our brain selects already one-in-a-million data from the total information provided by the environment. Our civilization tends to generate a huge amount of artificial information, and we do not know how this interferes with internal brain noise.

A genetically wired *visual cortex would be contaminated by fluctuations*. In the realistic model of the cortex, however, internal noise was effectively dampened. Goldenfeld suggests that *evolution has selected for a particular network structure that inhibits hallucinatory patterns*. The dominant patterns that spread through the network will be based on external stimuli—a very good thing for survival.

> If the cortex has been built with the long-range inhibitory connections all over the place, then the tendency to form these patterns would be stronger than the tendency to process the visual input coming in. It would be a disaster, and we would never have survived" Thomas said. *Because the long-range connections are sparse, the models don't produce spontaneous patterns,* unless you force t hem to, by simulating the effects of hallucinogenic drugs.

Experiments have shown that hallucinogens like LSD appear to disrupt the normal filtering mechanisms the brain employs, perhaps boosting long-range inhibitory connections and therefore permitting random signals to amplify in a stochastic Turing effect.

As these scientists showed, the hallucinogens disrupt the work of the brain filtering mechanism that inhibits long-range connections; a previous study indicated that the hallucinogens affect the entropy of the brain, lowering the activity in frontal lobes while enhancing the activity in other areas of the brain, which are normally less used. As it seems, a similar process is played by meditation.

> There are many higher-level hallucinations than Kluver's simple form constants. But when higher cognitive functions kick in, such as memory, you start to see more complex hallucinations, and you try to make sense of them. But they are just stored memories.

Independently of the paper mentioned above, Katy Waldman in her paper made an introspection in recent neuroscience research which papers came to at least two different conclusions. One of them says that "our ancestors found these stimuli aesthetically pleasing that the neurons are programmed to fire at the sight of them." However, the researchers into trance state have another theory. They argue that the patterns are not abstract at all. Instead, they are literal representations of the hallucinations that the artists experienced deep in the cave. Ironically, *the most abstract designs of the prehistoric era may have required the least symbolic thinking.*

The last conclusion changes even our current interpretation of "abstract" and "symbolic" thinking, meaning that these notions might have a hidden explanation in our oculomotor system.

I presented many quotes from those mentioned papers because these new knowledges provided here will help the reader to understand how the language evolved in the human brain, how the arts, symbols, writing, math, and logic also appeared out of the same neuronal structures. But the most important fact is that this shows how the new mechanism (mentioned in my paper), being developed in the oculomotor system, works.

Now, I will introduce another important paper for the subject of my work. It belongs to Michael J. Winkelman who teaches at Arizona State University and was published in September 2017 in *Frontiers in Neuroscience* under the title *The Mechanism of Psychedelic Visionary Experiences: Hypotheses from Evolutionary Psychology*. I will quote some ideas.

> Neuropharmacological effects of psychedelics have profound cognitive, emotional, and social effects that inspired the development of spirituality, of cultures and religions worldwide.

> Humans have a long evolutionary relationship with psychedelics, a consequence of psychedelics' selective effects for human cognitive abilities, exemplified in the information-rich visionary experiences.

Converging forms of evidence support the hypothesis that the source of psychedelic experiences involves the emergence of these innate cognitive processes of lower brain systems, with visionary experiences resulting from the activation of innate processes based in the mirror neurons system.

The role of psychedelics bind to human serotonergic receptors with a higher affinity than they do to those receptor systems in other primates. The interaction of psychedelics with the innate structures of the human brain produces novel forms of information and integrative cognitive processes. It suggests that **psychedelic substances** operated as environmental factors, *selecting for enhanced capacity for specific forms of information processing*.

Associated with the information presented above, it would be some important facts about the **transcranial magnetic stimulation** (TMS) that induces pulses of currents in the neurons. When TMS field is focused on the visual cortex, the subject sees lights which appear as discs and lines. Moving the field and the lights will move, too.

The same process can be repeated in labs with bigger magnets. Joseph Peer and Alexander Kendl of University of Innsbruckin Austria have calculated that *rapidly changing fields*, when associated with repeated lightning strikes, *are powerful enough to cause various types of visual hallucinations*, like luminous lines and balls that appear to float in space.

However, there are numerous reports of such types of observations during thunderstorms which storms play a similar role as in the TMS procedures.

As another domain with similar results on our brain systems is the domain of radiation, being it cosmogenic, chemotherapy, radiotherapy, hallucinogenic substances, or even the background high-level radiation.

Therefore, I would say that in nature are various sources, which can trigger the significant but uncommon activity of the human brain. *One of the major sources that probably developed unknown effects on human brain activity would be the geomagnetic excursions* where some rapid flips of the geomagnetic field occur. Such geomagnetic events extend over periods longer than one hundred years, affecting at least five consecutive generations of people and other species.

In this paper, I support the idea that our civilization was created by the works of a new mechanism associated to the oculomotor system. This mechanism would have the role in providing to our conscious states the information described at the beginning of this preamble that would be provided only from an unconscious state.

This process of new unconscious information generation to our normal awareness appeared at sudden, thirty thousand years ago, and since then it evolved in alternating sharps and gradual phases.

The fact that humans encountered mental problems during the Holocene would be well documented by the first skulls with trepanation discovered in France and dated as early as 6500 BCE. At that particular archaeological site from 120 skulls found as many as 40 had trepanation holes.

In general, more than 1,500 trephined skulls were found from the Neolithic period, and this amounts for 5-10 percent of all the Neolithic inventory.

Most researchers believe that trepanation was applied to subjects who complain of hallucinations and was affected by different types of mental illnesses.

THE SENSE OF UNCONSCIOUS AND CONSCIOUS PERCEPTION

The neuroscience experiments demonstrate that stimuli are perceived when subjects are not consciously aware of the stimuli. These experiments have concluded that conscious and unconscious processes lead to different qualitative consequences. All experiments have dealt with the duration of exposure to the stimuli. When the duration is shorter than 50ms, it has been categorized to produce unconscious perception. Now, when a stimulus is unconsciously perceived show a predominance of meanings (semantics), while the conscious perception discloses structural (surface details) characteristics. In short, unconscious is holistic and contextual, while the conscious disclosed the content.

Here, it also intervenes the impact on the produced type of memory. In the case of the unconscious perception, the memory of the perceived information stays longer than few seconds and even can produce a memory that lasts up to twenty-four hours.

These experiments indicate various domains of the unconscious, and this differentiation is related to the duration of the stimulus that generates a particular duration of the memory.

The duration of a stimulus is information about the stimulus regularity that is, in fact, its capacity to preserve unchanged the basic characteristics for a certain period. Shorter duration stimuli show that the characteristics remain unchanged only during this described time interval.

By contrast, conscious perception selects from all environmental information provided to the oculomotor system only that information that remains unchanged for at least 50ms. On average, the conscious selections are defined with a duration of 190ms; this means, such information remains unchanged for the mentioned time interval.

Now, the length of the lasting memory is a consequence of the same neural mechanism: things that change too fast (in less than 50ms) do not need to stay longer in the brain because they change and show no further utility for the brain processing. The conscious selection that only retains one piece of information out of one million pieces provided can use these selected pieces for a long time processing. Hence, the neural network is interested in generating long lasted memories, which are useful for a long time interval.

These experiments help us understand the mechanism that filtrates the information from the environment for the processing of conscious and unconscious processing networks.

In general, the filtration deals with an intrinsic capacity of information to change. When a longer duration (more than 50ms) is needed to change the information properties, this type characterizes the information that is selected for conscious processing. In reverse, a short time duration for change (less than 50ms) characterizes the type selected by the unconscious neural network.

However, the unconscious seems to have several types of filters, all of them based on various durations before a change can occur; but all of them are changes occurring in less than 50ms. Here, probably there are further separations based on duration. I can assume that that duration closer to less than 50ms are those selections which can be transferred into the conscious selection, or at least, they can influence the conscious selection, especially when we deal with decision-making processes. Consequently, the neural network generates memories with various durations based on their utility to neural processing. Eventually, there would be memories of less than a few seconds or even less than one second, which obviously would not be perceived but they will have a certain but unknown role in processing.

I would say that conscious selects three pieces of information out of a million, and unconscious also selects some other pieces from that same one million. It may amount to a total of 5 percent or so. The rest of the information shows changes that occur in less than one second. This rest of information defines what we call chaos.

The slow changing information selected by conscious generates our world of order that can be predicted and analyzed and that allows mental (so-called mathematical) computation. We call it periodic because it repeats itself or it shows a pattern that repeats itself; it shows symmetry, linearity, proportionality, and is finite.

The rest of information provided by the environment have opposite characteristics named chaotic: nonlinearity, asymmetry, non-proportionality, and infinity (no repetition of patterns).

However, later in this paper, one will see that several scientists defined consciousness and the brain in general as a massive reduction in entropy; according to some estimates, the brain entropy is reduced some one million times.

I have to mention that entropy, as a rule of Thermodynamics Second Law, always increases but can decrease locally, as in the case of the complex structure of life; the local decrease pays for this decrease with a bigger increase outside of that locality.

According to such, a view the conscious state seems to be low entropy. But this state defines an overall state that is called macrostate. Multiple microstates form this macrostate.

Recent research found that when more deep in sleep the mind is, the lower the entropy. But here, we probably deal with a certain decrease in entropy for several microstates, while the scientists did not found where elsewhere the entropy increased.

The same research found that more awake the mind is the higher the information content, the higher the number of neural interactions, the higher the values of entropy are.

In my opinion, we deal here with a variety of multiple microstates, and the researchers have been able to measure only local groups of certain microstates.

Overall, the same research estimated that *conscious states have higher entropy than unconscious states*, and this situation indicates an increase in connectivity for the conscious state.

A similar case occurs during meditation practices: the local entropy decreases in the frontal lobes and increases in other distant areas of the brain which distant areas at present typically show a low local entropy and seem to provide unconscious information. Hence, as appears, the meditation produces a redistribution of entropy all over the braingeneration a re-connection of the old neural circuits.

The same process of re-distribution seems to indicate that *historically we deal with a gradual increase in entropy in the area of the neocortex and this fact is related to the evolution of the conscious processes in the last several thousands of years.*

Guy E. Blelloch of Carnegie Mellon University, in a paper titled *Introduction to Data Compression* (January 31, 2013), explains:

> In particular a system is assumed to have a set of possible states it can be in, and at a given time there is a probability distribution over those states. The definition of entropy indicates that the **more even the probabilities, the higher the entropy (disorder) and the more biased**

> **the probabilities, the lower the entropy**. In the context of information theory, Shannon simply replaces" state" with "message." The self-information represents the number of bits of information contained in it and, roughly speaking, the number of bits we should use to encode that message. The definition of self-information indicates that the *messages with higher probability will contain less information.*
>
> The entropy is then simply a probability-weighted average of the self- information of each message, that is the average number of bits of information contained in a message picked at random from the probability distribution."
>
> *Larger entropies represent larger average information, and the more random a set of messages (the more even the probabilities), the more information they contain on average*.

From the above quotes, taken from Dr. Guy E. Blelloch's paper, one understands that the *conscious uses lower probability, fewer information data acquired by the oculomotor system from the surrounding environment.*

By contrast, the *unconscious uses higher probability, more information data that assures a hierarchical distribution, and where unconscious has the lead.*

Here I like to recall the issue of emotions and emotional thinking because it relates to probabilities.

There is an evolutionary path that might suggest that systems underlying basic affective reactions originated before systems for conscious awareness. This basic was the experience with the environment which experience was the generator of emotions as relatively short-term processes and conditions associated with the positive or negative experience. This experience leads to a manifestation of action as impulses.

The emotions were a primitive and less effective way of thinking. The experience with the environment was contextual because it had to validate with all senses combined the multitude of environmental causes producing certain effects. The effects were perceived holistically, meaning they were a variation from one pole to another, covering everything from a good/positive experience to the bad/ negative one. This type of mental introspection led to the meaning of quality in itself: positive and negative, and established the general polarization of the thinking patterns. Here, some scientists see a relation between emotions and a genetic fixed content/condition.

No feeling is permanent, constantly oscillating from positive to negative. *Emotions are constantly changing, showing their origin into an information content with larger variability or randomness than the one we experience during the conscious processing*. They run the entire spectrum of possibilities, which are probabilities.

Some experiments that connected the internal level of emotional stress with heart rate variability (HRV) found that the decision-making process leads to a large change in HRV. This result indicates the magnitude of emotional value connected to external stress that is a genetic dependency, too. It might stand for the origin of language-genetic-dependency that is the essence of a primordial emotional tongue.

It is widely accepted that cognitive processes and states can be unconscious, when occurring below awareness, or implicit when occurring without attention and intention. Here, the implicit is caused by a change in experience. By contrast, explicit emotion refers to a person's conscious awareness.

One can say that our Paleolithic ancestors had an emotional thinking system that was contextual, their intelligence was based on "understanding wholes," and their life experience led to evidentiate the quality of things.

As John A. Bargh and Ezequiel Morsella said: *"both phylogeny and ontogeny actions of unconscious mind preceded the arrival of a conscious mind, while here the action precedes reflection."*

The prehistoric mind showed a long-term adaptation to the ecosystemic needs.

HOW A MUTATION IN THE THINKING PATTERN INFLUENCED THE EVOLUTION OF LANGUAGE IN THE EAST AND IN THE WEST

It is interesting to investigate how the evolution of language has changed the way people in Asia thought differently than people in Europe. This distinction left a strong imprint on the cultural evolution of these regions that are still well preserved in our modern world. Before starting to analyze the distinction mentioned above, I have to mention some work that defines first the origin of imagination in the emotions.

Adam Morton, a philosophy professor at the University of British Columbia, explains:

> An emotion is a state we can enter into where is generated towards our having certain kinds of thoughts centered on a particular theme. Emotions must be directed towards a specific person or object-sharply distinguishing them from moods, which are more general collections of characteristic feelings that we move between. Imagination is the faculty of mentally representing a possibility in a purposeful way. *Emotion involves the capacity for imagination* because the emotion

> is a kind of state where we experience pressure towards generating imaginative representations of possibilities about some specific object.

As it derives, the emotional thinking was the formative element that shaped the imagination and an adopted imaginary thinking. The process of imaginary thinking was developing starting thirty thousand years ago, but it took a very long time until language itself turned split toward two distinct cultural paths: the *concrete language*, where the things can excite the senses (they can be sensed), and the *abstract language*, where evolved an abstract imaginary perspective.

These mentioned distinct paths, being taken by the imaginary thinking, seem to have a genetic background represented by the dispersion of Y-DNA and the R1b haplogroup from Western Urals to the Caucasus, while it entered Europe through Anatolia and Greece (in the second millennium BCE).

Analysis of Modern Greek and Cretan lineages point to a Neolithic, Bronze Age, expansion of the haplogroup R1b1a1a2a2-Z1203 lineage and R1b1a1a2-M269 lineage. The same genetic element is found in the same era in Italy, Sicily, and Sardinia. It is known that, in general, R1b haplogroup represents today the most extensive lineage all over Western Europe.

The mentioned R1b lineages are the most extensive in Minoan samples from Crete and Mycenaean samples from mainland southern Greece, but also in the population of Attica that was the region that circumscribed ancient Athens.

Athens is the origin of Western philosophy and thought, and the same culture was gradually adopted all over Western and Southern Europe. During the Hellenistic period, this philosophy spread into the Middle East and North Africa.

This differentiation between the new and the old thinking paths is clearly expressed in language but also in the local cultural traditions of the west as well as the east. I will describe only several linguistic differences.

Eastern thinking that was the old ancestral thinking developed a "concrete" language: a language where the things can strike the senses, where the things can be seen and touched, and generating emotions. The 3D environment deeply influenced this "concrete" perspective. It led to a type of "block logic" that represents blocks of parallels, where there is no chronological sense, and the things are ordered according to the action's needs and purpose.

Another element was the representation of time. For example, in biblical Hebrew, the past is in front, while the future is in back. In the Eastern languages, all nouns and verbs are dynamical.

The most summarizing characteristic is about the way the East see the world holistically and contextually.

In opposition to all these, the R1b Athenian mind came out, as early as the 8th Century BCE, with a very different perspective that is called the "abstract or imaginary perspective" and generated a type of "abstract or imaginary language." This abstraction or imaginary represents "the linearly thinking mind."

They believe that the future should be in front and past in back, but this new approach took centuries of education to be properly implemented.

Not coincidently, both Minoans and Mycenaean developed the Linear B writing. Their nouns were static, while the verbs were dynamic.

This "abstraction" introduced the chronology of time that is displayed in a linear manner that is called "step logic." Many sources insist that the Hebrew Bible was written in a "concrete" perspective, but its translations changed into an "abstract" perspective.

This "abstract thinking" of the Attica people started to break wholes into parts for analyzing them. From this originated the current meaning of "logic" that leads to analysis and synthesis. It is a thought process that tries to search for the "content" of things, while mostly ignoring the context.

However, here was the ancient linguistic assumption that the meaning of an expression is built up from the meaning of its parts and the way they are put together. But this approach led to a vagueness, ambiguity, and fuzziness of the meaning; and it was hiding the true origin of every true meaning, giving way to that abstract, imaginary world where everything is interchangeable while being based on straightforward, direct causality. Most of the world culture adopted this view, making us live with such damaged chimera.

Sartre (1940) referred to an image as a form of consciousness. Lancan, in response to Sartre, presented the imagery as one of the three intersecting orders that structure all human existence, the others being symbolic and the real.

Wikipedia explains:

> The imagery as a Lancanian term refers to an illusion and fascination with an image of the body as coherent unity, deriving from the dual relationship between ego and the specular of mirror image. The illusion of coherence, control, and totality is by no means unnecessary or inconsequential (as something that is illusory). *The term imaginary is cognate with fictive, but in its Lancanian sense it is not simply synonymous with fictional or unreal; or the contrary, the imaginary identifications can have very real effects.*
>
> *The social imaginary is nevertheless an institution in as it represents the system of meanings that govern a given social structure.* The imageries are to be understood as historical constructs defined

> by the interactions of subjects in society. In that sense, the imaginary is not necessarily "real" as it is an imagined concept contingent on the imagination of a particular social subject."

The interplay between emotions and reason, desire, the symbolic order, and the real are representational modes of imaging that have both facilitated scientific developments and sometimes misdirected a sense of objectivity and certitude.

Technological meaning becomes located within larger social imagery.

There is reasonab e to assume that such a drastic change in the thinking path was not a random fact, but it was driven by one or several mutations in the brain, 7,000-4,500 years ago when the Solovki Geomagnetic excursion was active especially over Eastern Europe, Urals, Caucasus, and Anatolia.

However, I like to cite a paper titled *The Role of Culture in the Acquisition of Nouns and Verbs* published in Oxford University Press by Tracy A. Lavin of Northwestern University, D. Geoffrey Hall of the University of British Columbia, and Sandra R. Waxman of Northwestern University. The authors indicate that "...children tend to emphasize nouns during book reading sessions, and verbs during play sessions."

> Recent work in cultural psychology suggests that in different cultures, social cues may emphasize different sources of information and, in particular, some cultures may highlight information that supports noun learning and others may highlight information that supports verb learning. Members of Eastern cultures tend to engage preferentially in a holistic type of reasoning, while members of Western cultures tend to engage in more analytic reasoning. Members of Western cultures may be

> particularly attentive to sources of information that are helpful in acquiring nouns, while members of Eastern cultures, as a consequence of the holistic reasoning style, may be more attentive to verb-relevant sources of information.
>
> Easterners have shown a preference for relational categories, while Westerners show a preference for taxonomic categories. For example, Chinese participants are more likely to chose the monkey, and the banana (focusing on relation "eat") and Americans are more likely to chose the monkey and panda (forming a taxonomic category like mammals or animals). In memory tasks, *Easterners seem to be more attentive to relationships between objects themselves,* especially salient focal objects. For example, Masud and Nisbett showed to Japanese and American subjects animated underwater scenes, and later asked participants to describe what they had seen. The Japanese made more reference than the Americans to relationships between different objects in the scene and between objects and the background. The Americans focused more on describing individual objects.
>
> *The differences in focus (that is attention) could be relevant to the acquisition of nouns and verbs.*

In the meantime, Nitya Sethuraman and Linda B. Smith of University of Michigan-Dearborn (January 2012) indicated that:

> studies that have examined whether English speakers can figure out the verb from watching

> a scene with minimal contextual information suggest that linguistic context, particularly *the argument structure of the verb, plays a strong role in isolating the relevant meaning*. Young English learners find the range of argument structures with a verb highly informative regarding important aspects of that verb's meaning.
>
> All this evidence suggests a strong relationship between expressed arguments, a verb meaning, and the relational structure of referred-to-scene. However, not all languages present such strong links in everyday language between these three components, in that many languages do not explicitly mention the verb's arguments.

Here, one may wonder why the humans split so sharply in a matter of thinking modes? The explanation refers to evolution itself.

The Paleolithic people were holistic, contextual, had a visual/nonlinear thinking, and all these things have been produced by several million years of adaptation and experience. The revolution in evolution sometimes came between forty thousand and thirty thousand years ago, and it was materialized by rapid while sharp advances in language, where the introduction of verbs was the paramount part of that revolution.

One element of that revolution was an increase in human attention. As we have seen, still **humanity lives with a split in the way attention is divided between nouns and verbs**. *The Western culture remained tributary to the Paleolithic attention for nouns, while in opposition to this approach the Eastern culture turned the attention to verbs.*

However, this change in attention did not reflect the final result obtained because here the Eastern culture appears as the keepers of the old way of thinking that was visual, nonlinear, holistic and contextual. The Eastern thinking adopted the dynamic virtual space of the mind introduced by verbs, even when the verbs are not well expressed in the languages that belong to the Eastern culture.

As appears, the Western culture adopted more of the mechanism regulating the language functions, including its linearity, while emphasizing the objects presented in the form of nouns.

This combination of paths led the *Westerners to treat the reality as a language, or one can say that they turned to think in the same way they talked.* In other words, *language became thinking*, but this was mostly linked to the language of the nouns, for which they pay maximal attention. The noun remained the subject, while those subjects became the objects of decompositions (analysis) and recomposition (synthesis), which was the way they treat their everyday language.

The verbs remained the elements which allow the maneuvering.

The process of adapting thinking to the language inner rules (which they eventually were genetic in fundament) *has split the human evolution into two main groups of adherents: the followers of the nouns in the West, and the followers of the verbs in the East.* The nouns were "abstract," but the verbs represent a "concrete" approach vindicated by way of action.

However, *the "abstract" allowed huge maneuvering that led to mentally produced abstract ideas, while stimulating further a material manufacturing of such "abstract" ideas;* on the converse, the "concrete of the Easterners" brought a "realistic" limitation about altering the reality. *In the East, the reality is solid and cannot be abstracted in any easy material ways; it cannot be altered because it cannot be broken into parts.*

In the East, when one ponders, one must first eliminate this solid reality from the thought, and only secondly, the mind tries to reach the "abstract." The eastern "abstract" is fundamentally different from the western "abstract."

Considering the illusory nature of reality, there is no stimuli or no justification in the East for one to try to alter it. For ages, in the east, everything was done at a minimal level of materiality or to assure the subsistence.

In the West, the common reality originates into an "abstract" meaning or sense, but here no one likes to acknowledge that the illusion itself produces what would be called the "illusory material life" or an "illusory civilization." Taking the "abstract world" as fundamental, one has only to justify its "undeniable materiality."

In the West "conscious things" are part of the "material world." In the East, conscious things are the illusion, and the unconscious is the "abstract."

Today, most researchers have validated that conscious is a selection of one-in-a-million pieces of information (or bits), which are those bearing lesser information. Is this "selection" an illusion or only an incompleteness? Is incompleteness an illusion?

However, recent research shows the current state of nonlinear science in psychology, particularly relevant in the areas of emotional dynamics and resilience. Most of nonlinear refer to disproportional cause and effect.

Various experiments on emotional dynamics found the subjects displaying continuous fluctuations, let's say, from happier to sadder over a short period, like in a two-and-a-half-minute period, where have been measured four hundred fluctuations. They found evidence that the fluctuations were characteristic of low dimensional chaos because they were both deterministic but also unpredictable. They found them to be fractal, while happiness was found to be a self-organizing process. In the meantime, happiness is a simple scalar variable, and individuals respond consistently and mechanically to that variable.

A higher entropy was associated with measures of positive emotional states. In arousal patterns being detected by nonlinear methodology, have been found five hundred continuous shifts in the process of self-evaluating with a duration of four minutes.

All examples mentioned above show how nonlinear is the human emotional dynamics that stays in deep contrast with the linearity of logic that is the linearity brought by linear language.

Even then, the Easterners still preserve more nonlinearity that the Westerners because their languages help preserve nonlinear attributes, like higher emotional dynamics. The Eastern use of verb-dynamics expresses these dynamics.

HOW HUMAN MIND INVENTED THE CONCEPT OF THE FUTURE

As appears, during the Paleolithic, humans use to place the past in front and the future in back. It is how current primitive people deal with past and future. Also, they employ an environmental concept in localizing things, like up and down, or based on cardinals, or as right and left. I will call this a 3D expression.

As the archaeological research and discoveries indicate, humans and humanoids had the concept of life after death. We know this fact because we found burial sites where the prehistoric individuals were put to rest with their belongings and other things, which were assumed to be needed in their life after death.

We also know from archaeological evidence that local shamans have been in charge of predicting future outcomes.

The old traditions talk about the interpretation of dreams for predicting the future that, during the Greek antiquity, was named Oneiromancy. In a more sophisticated way, this took the form of the institution of Oracles.

As we have seen in the previous chapters, the use of hallucinogenic substances has produced special states of the mind where some individuals were making predictions outcomes to occur.

I have to mention an early concept about destiny, where the future of everyone was well defined.

All these traditions indicate how gradually the idea about "future" evolved within the human mind. The introduction of tenses for nouns and verbs had a major contribution in shaping further the concept of "future."

Even then, we have massive evidence that almost until recently or even at present, some people continue to place the past in front and the future in back.

Many languages have various grammatical ways in which the future is expressed as a matter of the past that continues in the present and other variations that have the future as the past.

However, it existed a prehistoric common sense that all predictions are based on memories and all memories have dealt with the past. Even the dreams were seen as another form of surfacing memories.

The science says the future is in front and shows asymmetry compared with the past; thus, the future is not like the past. The most paramount proof offered by science is the general entropy of the universe that shows that the entropy always increases, implying that time flows only toward the future. It is called the arrow of time.

Now, as I mentioned before in this paper, the biological systems are characterized by negentropy, which means that the general entropy can be reduced locally, while to compensate, it must increase in the surrounding space or elsewhere. This negentropy still respects the provisions of the Second Law of Thermodynamics.

However, in this particular case of the biological systems, one can assume that the time would flow in reverse that is from present to the past because the entropy is reduced, diminished, and hence it flows in reverse, too. This local feature is explicitly represented by the works of our oculomotor system where the prediction this system makes is about the past, implying a flow from present to the past.

This argument of mine presupposes that in the universe there are two types of flows of time, and they act in opposition one to another. When biological systems mature, they indicate an increase in entropy that represents their tendency toward disorder, leading to termination of existence.

When human attention increases, one begins to observe things from the pastwhich past unobserved. And then, one tends to wish to experience those things that past unobserved. But these novelties still are the things of the past. This desire contributes to the creation of a time where those somehow novel experiences can be fulfilled. It is an abstract time that gradually gained consistency in the human mind when the attention span increased. In a way, this abstract time became asymmetric because it included previously unobserved things. Is this abstract time what we have called the time of future?

From the view of entropy, the *future is that particular time when the things turn increasingly disordered*, causing the dismantling of the established or ordered entities. The *future reflects an increase in entropy, while time in general represents an entropy change.*

Here, we do not perceive time because of brain entropy is increasing or decreasing, but because of the lowness (low-level) it reaches.

Is entropy another perceived process, another mental construct? As it seems, entropy is not a physical process in itself, but a similar mind-construct as the time-construct is.

Many elements that form one system evolve on different time-scales, and there is a wide-spanning of time-scales between any cause and any outcome because there are many degrees of freedom between which takes place the transfer of energy. Hence, a system shows many types of entropy at work.

Are unconscious and conscious systems? If they act as systems and probably they do, then we should have various types or degrees of entropy working instantly in our mind.

The researchers at the University of Wisconsin-Madison discovered that "*visual past images* are not stored somewhere in the brain, but *get actively reconstructed*." This study indicates that *the imaginary thinking benefits from a "reconstruction" of the past recorded images in the memory. It shows that the future is such an imaginary reconstruction that, in the end, has nothing to do with the experienced past.*

PERCEPTION IS A HYPOTHESIS: RESHAPING OF REALITY INDUCED BY THE MIND

Dr. Patrick Cavanagh with University Paris Descartes in France and Harvard University, USA in an article *Visual Cognition* published on July 1, 2012, explains:

> A critical component of vision is the creation of visual entities, representations of surfaces and objects that do not change the base data of the visual scene but change which parts we see as belonging together and how they are arrayed in depth. The entity that is constructed is unified in our mind even if not in the image.
>
> In almost all cases, each construct is a choice among an infinity of possibilities, chosen based on likelihood, bias, or a whim, but chosen by rejecting other valid competitors. The entities are not limited to static surfaces or structures but also include dynamic structures that only emerge over time - from dots that appear to be walking like a human or a moon orbiting a planet, to the causality and intention seen in the interaction of dots, and the syntax and semantics

> of entire events. There is clearly some large-scale information processing system that accumulates and oversees these visual computations.
>
> The descriptions of surfaces, objects, and events are not solely for consumption in the visual system but exist at a level that is appropriate for passing on to other brain centers. Clearly, the description of a visual scene cannot be sent in its entirety, like a picture or movie, to other centers as that would require that each of them have their own visual system to decode the description. The nature of this high-level, visual description that can be exported to and understood by other centers is as yet, completely unknown.
>
> Most vision scientists avoid being too explicit when presenting concepts about visual cognition, having learned that explicit models invite easy criticism. What we see in the literature is ample evidence for visual cognition, but few or only cautious attempts to detail how it might work.

David Vernon with the Institute for Cognitive Systems, Technical University in Munchen, Germany, wrote in a paper *Visual Cognition*, explains:

> Visual cognition constructs visual entities by collecting perceived parts into coherent wholes, determining which parts belong together. Since the sensory data on which the processes of visual cognition operate are typically incomplete and insufficient to specify the percept of which we

are aware, there are many possible solutions or interpretations. Consequently, additional extra-retinal information often referred to as object information, is used by visual cognition to infer what the percept is.

The dynamics of the visual entities can be used to infer a causal relationship between the events or to attribute some sense of intentionality to the entity. Thus, the unconscious interference of visual cognition also impacts on the construction of a theory of mind. A key role of representations of visual cognition is their use to communicate with other centers of the brain. Thus, visual cognition provides a bridge to general high-level cognitive function while still making its own independent cognitive interference. In essence, visual cognition constructs a representation of the visual world which is constantly updated on the basis of new visual data.

The processes of visual cognition are held to be principally unconscious, operating rapidly on the flux of visual data sensed by the retina in order to choose the conscious percept of which we become aware.

The issue revolves around the cognitive impenetrability of visual perception: whether or not any cognitive functionality such as inference or rationality is involved in visual perception. The visual system does have its own independent cognitive processes, quite apart from the more

> general cognition that occurs in other parts of the brain and with which the processes of visual cognition interact.

Extra-retinal signals are non visual signals used by the central nervous system to encode the position of the eyes in orbit. The brain combines these signals with retinal (visual) signals. The relative contribution of efferent and afferent information to extra-retinal signals is still highly debated. (Encyclopedia of Neuroscience, 2009)

In my opinion, the above definition of extra-retinal information is incomplete because it implies the work of neuronal systems. For this reason, I will call it "contextual," while the retinal information refers strictly to the visual "content."

We will see later in this paper what I suggest about the works of these contextual and content-related systems.

However, also in my opinion, here intervenes the quantum effect of the observer over the essence of reality. In other words, the design that is made for neurological experiments influences their result, hiding the "true loading" of latency that is induced by neuronal processes. In short, it hides the probabilities or options embedded in that loading.

This idea appears, among other authors, in Z. Pylyshyn's 1999 paper *Is vision Continuous with Cognition? The Case for Cognitive Impenetrability of Visual Perception*. He defends the position that "an important part of visual perception is prohibited from accessing relevant expectations, knowledge, and utilities in determining the function it computes—in other words, it is cognitively impenetrable."

The paper shows that "vision" is "intelligent" and involves elements of problem-solving. The paper also examines some examples where instructions and hints are alleged to affect what is seen. The evidence is more readily assimilated to the view that when cognitive effects are found, they have a locus outside early vision.

The author indicates that vision is more than seeing because it involves "intelligent processes, which for example, cannot be defined in experiments. Cognitive effects, which load the vision, are outside the vision In scientific experiments, the instructions and other elements of such designs like hints affect what is seen."

SUBLIMINAL AND UNCONSCIOUS LOAD INVOLVED IN LATENCY

Latency load is comprised of the "possibilities" which are "probabilities," but which represent information about reality that we are not aware of while being ready to exploit it unconsciously.

Jan Theeuwens with the University of Amsterdam published in 2009 research funded by a grant from the Human Frontier Science Program.

> This thesis examines the mechanism underlying visual-spatial attentional selection. The research demonstrated that top-down processes in spatial selection are voluntary, intentional and goal-related. The bottom-up processes are involuntary, unintentional and unrelated to a goal. Selection can occur with or without the eyes. Research has shown that a highly salient object may get selected without consciously perceiving it.
>
> The goal of this study was to investigate whether subliminal information influences basic oculomotor behavior. A visual stimulus that is not consciously perceived may be able to affect the oculomotor processes, and it involves the superior colliculus.

This research comprised several experiments divided into distinct sections for interpretation.

Chapter 2 concluded that "subliminal cues can capture attention in a bottom-up fashion. It suggested that a direct subcortical route from the retina to mechanisms essential for attention (retinotectal pathway), such as the superior colliculus, was responsible for unconscious attentional orienting."

> It was investigated whether a subliminal distractor affected eye-movement metrics, like saccade trajectories and endpoints. Deviation of saccade trajectories is known to arise from competition between a target and distractor related activity within the oculomotor system. Results suggest that subliminal visual information evokes competition in the oculomotor system."
>
> "We conclude that unconsciously processed spatial cues can affect the oculomotor system and provide evidence that the retinotectal pathway may mediate unconscious attentional processing."
>
> "Saccadic target selection is initially based on bottom-up salience information. After the initial bottom-up activity, top-down influences can enhance or inhibit this activation."

The quotes presented in the above paragraphs describe the processing of subliminal and unconscious information in the oculomotor system by the intermediation of latencies. So, all non visual information involved is provided by the brain's neuronal systems and is unconscious.

THE HIDDEN CENSOR

On June 11, 2013, Professor Keith Payne of the University of North Carolina at Chapel Hill published a very interesting article in *Scientific American*.

He wrote that psychologist, Ulric Neisser demonstrated in 1980 that

> ...when attention is occupied with one thing, people often fail to notice other things right before their eyes." "The key is to realize that this is just what attention is: selectivity. For a brain with finite computing power, zooming in to focus on one thing always means picking up less information about everything else. That's how we are able to concentrate on anything at all and leave behind the blooming buzzing bundle of distraction that is the rest of the world. Prioritizing one thing and neglecting everything else are two sides of the same coin. We miss some unattended things more than others.
>
> In new research, my colleagues Jazmin Brown-Iannuzzi, Sophie Trawalter, Kelly Hoffman and I pushed the idea of selective selectivity further by asking whether the unconscious screener might have priorities of its own. Scads of studies have suggested that the unconscious mind is riddled

with stereotypes and biases, even among people who are consciously well-intentioned. We asked whether the unconscious screener is prejudiced. As we suspected, who they saw depended on what they had on their minds. The simple fact of selectivity has big consequences: at any moment we are aware of just a tiny fragment of all that is around us. Our conscious experience seems to be a rich and detailed picture of the world because where it's not, we aren't paying attention.

The idea of selective selectivity means that the unconscious mind may be shaping our experience even more dynamically than previously thought, screening what we see based on goals and emotions.

This makes good evolutionary sense. It is important to be selective so that the mind can devote most of its resources to the task at hand. When it paid to see a number, people saw a number. When it paid to see a letter, they saw a letter. Some events become visible or invisible based on what they mean to you.

The unconscious screener must be able to do at least two things. First, it has to know what the goal is. Second, it must make a first approximation of whether the candidate for consciousness fits the goal or not. At a minimum, our findings imply that the unconscious can represent social goals, and seems to have "opinions" about which is suitable to which or to whom.

> The unconscious always has a head start. The power of the unconscious is greatest when our attention is under the heaviest demands.
>
> In today's multitasking world, we delegate even more to the unconscious. It makes you wonder who you have looked at today and have not seen.

Payne's article provokes further questions. One of them is about the experimentally proven existence of multiple levels of the unconscious which act as a filter for the information sent to the conscious level. This filter operates selectively. Additionally, why do other animals not show similar selectivity?

I propose this is due to speech-generated latency. This latency is created only by speaking, and it does not occur in non-speaking animals. This latency is like a door or window created at our conscious level and allows the unconscious selections to penetrate the conscious level. When this door or window does not exist, there is nothing to come into the consciousness beyond the normal workings of the visual system.

The window above, being opened by the speaking process, is unique to humans in the animal world and is the sole provider of unconscious information that is not intended to fill a gap in exogenous motions. This unique case allows the unconscious to divert its processing and filtering toward something else of evolutionary importance. This new type of processing can be modulated to fulfill goals which are intentions. This unconscious information assists human intentions, helping to find things, to discover, and to innovate.

Human evolution benefited from this new mechanism and associated abilities, which both appeared some forty thousand years ago. However, the case of primitive people of our time seems to indicate that the mentioned door or window has appeared in connection with some geomagnetic events, only affecting people in certain geographic areas and not others.

A CLIMATE PULSATION MADE HUMANS LEAVE AFRICA

Most of the general knowledge literature avoids specifying the geographic, geomorphological and climate conditions which together made humans leave the African continent in several migration waves.

There is known that between 185,000 years ago and around 70,000 years ago North Africa (including Sahara) has experienced a wet period with short phases of dryer climate. It is the time when the Homo sapiens genetics have been set in place.

I like to mention several dry episodes of this generally wet period: they have been determined mostly in South Negev Desert (Israel), but it is generally accepted that they bear similar conditions all over North Africa. The dry spell was dominant in the era 110-90,000 years ago. This short episode of drying up corresponds with the first Homo sapiens adventure into the Levant. The wet climate then returned from ninety to seventy thousand years ago, and the sapiens returned to Africa.

Around seventy to sixty thousand years ago, the climate turned to arid and in some locations, it became very arid. This era extends to almost thirty thousand years ago.

It is the climate moment (probably sixty thousand years ago) that made humans from the rapidly drying out regions to migrate out of those areas, crossing into the Arabian Peninsula and the Levant.

However, there is to consider that short-term drying out episodes were cyclic, and they eventually caused several other waves of migration toward the east (Eurasia).

The geological evidence indicates the end of Green Sahara around seventy-two thousand years ago. Since then, there have been several wet/humid phases: 65-61 ka, 52.5-50.5 ka, 37.5-33.5 ka, 15.5-5.0 ka. Obviously, between them have been a ride epochs.

In one of these dry (arid) phases, like one that started fifty thousand years ago, a large out of Africa migration occurred.

However, the out-of-Africa event requested more energy for the brain that was dealing now with increased "uncertainty" generated by new while unknown environments.

This event, in my opinion, contributed to the increasing of the brain size, which size, shown by the larger skulls, is between eighty thousand and twenty-five thousand years ago the largest on the archaeological record.

FROZEN-IN-TIME

A brief history of the status of primitive people and their languages as they exist today: The people of the Americas before Columbus were in various stages of advancement. The Incas of Peru and other peoples of Mezzo-America were in the advanced stages of metal manufacturing that started in the South around 2000 BCE and in Mezzo-America around 500 BCE. The people of Amazonia started an agricultural civilization around 6,000 years ago and reached their peak around 2,500 years ago. People all over the Americas were practicing agriculture and had settlements and stone and wood tools.

In the North they used copper; in the South they used iron.

In Amazonia, people domesticated to some degree at least eighty-three species of plant. Another three to five thousand plant species, which were not domesticated, were nevertheless used in some way.

Amazonia reverted to wilderness after Columbus' arrival because of European, African, and Asian diseases to which Native Americans had no immunity.

In the Southeast Australian region of Victoria, the Aboriginal people practiced agriculture, developed aquaculture where the climate permitted, built permanent settlements of stone, and had a long-distance trading network. In other regions of Australia, they did not practice agriculture, being hunter-gatherers.

People are inhabiting Papua, New Guinea had agriculture and some 850 languages.

The same can be said for the Khoisan people of Kalahari who remain hunter-gatherers today.

What kinds of languages do aboriginal peoples speak? Numerous Australian Aboriginal, Papuan, Amazonian, Sub-Saharan, and Siberian languages provide examples of a grammatical complexity that is not found in familiar European languages. As it appears, complexity in language is not a result of clever people developing the language as civilization emerges.

Professor Roberta Lindsay explained in her publication dated July 28, 2009:

> There is false claim out there that primitive people have primitive languages. That's completely wrong. The wildest, crazy, complex and indecipherable languages out there that we almost still can't figure out are the more primitive type of languages spoken by more primitive, isolated and less civilized or at least modernized groups.
>
> Once a group gets civilized or modernized, the structure of the language undergoes simplification, often massive simplification. The more civilized or modernized the language, the simpler and more dumbed-down the structure of the language is.

Miekko also explains:

> There are no primitive languages. All languages have a system of sounds, words, and sentences that can adequately communicate the content of the culture. The languages of the so-called

"primitive" people are often very complex in their grammatical structures. There seems to be no correlation between a language's grammatical complexity and the technological level of the society or other aspects of culture. However, cultures that have more complex, diverse economies and advanced technologies have larger vocabularies.

DEFREEZING OF TIME

The case of these primitives, reaching into our modern era, opens again the issue of specific mutations generated by the cosmogenic radiation penetrating the Earth's geomagnetic field. If not exactly mutations, they may have nonetheless represented important changes in gene expression.

What we have here is all "primitives" speaking a complex language. It indicates that when they migrated to their current locations, they were speaking a primitive but complex language. It means, at that particular time, all people on Earth were speaking languages. Did all these languages originate in the geomagnetic excursion occurring eighty to seventy thousand years ago or even before that time, originating in previous excursions? It shows that the geomagnetic excursions eighty thousand years ago and before had a global influence on humans.

The primitive people from Australia and New Guinea came to their current locations some forty thousand years ago. What happened to them in this forty-thousand-year interval?

They evolved from a Paleolithic type of society into a Mesolithic-Early Neolithic society, starting some form of agriculture, and settlements. It appears that the people in New Guinea were connected over the Torres Strait with Cape York Peninsula of Australia, and indirectly to Eastern Australia, going all the way down to Victoria, and these are the places where farming ignited. But these people remained separated from other primitives living in Central, Western, and Southern

Australia where hunter-gatherers preserved their original state of civilization unchanged.

We know that the path of these geomagnetic excursions has been detected in rocks in some locations but not in the same type of rock in other locations. For example, the Laschamp Excursion, roughly forty thousand years ago, occurred when the Cro-Magnons were already in Eastern and Central Europe. The path of the excursion crossed Europe, Africa to its East and South, went over the Indian Ocean to Tasmania, and up along Eastern Australia and New Guinea all the way to Siberia, reaching back to the initial location of the North Magnetic Pole.

Some other excursions in the era of roughly forty to twenty-five thousand years ago crossed from the North Magnetic Pole to Europe, taking various paths over the Atlantic to Mezzo-America and Peru, curving over the South Pacific Ocean until they reached Eastern Australia, Eastern Asia, and Northeast Siberia. It was known as the Lake Mungo Excursion.

The aboriginals of the Americas came here between fourteen thousand and twenty-eight thousand years ago. They, too, came with a Siberian spoken language. They incorporated the mutations produced by the entire group of excursions roughly twenty-five thousand to forty thousand years ago, being in a similar stage of Late Paleolithic civilization as most of Eurasia and Africa was at the time.

There was also the Hilina Pali Excursion, also known as the Tianchi Excursion of China seventeen thousand years ago and the Gothenburg Excursion thirteen to twelve thousand years ago, which both opened the way toward domestication of animals and plants and transited into the Mesolithic.

It appears that agriculture began in the Americas almost within the same time frame as Eurasia, which is eight to six thousand years ago. Local differences emerged, such as types of animals and plants domesticated in some locations but not others. Some people in the Americas settled east of the Andes, and some remained to the west.

Further geomagnetic excursions, occurring between 2,500 and 7,500 years ago, affected Eurasia, parts of Africa, Mezzo-America, and the Peru-Ecuador regions but not North America, Australia, New Guinea, or some other parts of Africa.

I could say that the Solovki Excursion of 4,500-7,500 years ago was responsible for the human transition toward the Bronze and Iron Ages and into Early Antiquity. However, this mutagenic effect appeared quite localized to the Eurasian part of the Northern hemisphere with a faint reflection in Mezzo America, and Ecuador-Peru.

The Etrussia Excursion of 2,700-2,500 years ago was mostly localized over the previous areas; it opened the way toward Classical Antiquity.

If the picture being drawn above about the mutagenic effects of cosmogenic radiation is correct, this would suggest the origins of the differences in human development at various geographic locations. The map of crisscrossing excursional paths was the map of "options." When more crossings occurred, there were more "options." The mutagenic change opened one window after another.

CHANGES IN GENE EXPRESSION

This mutagenic effect would be reflected in a new mechanism in the human brain. It is known in the literature that differential gene expression reveals that qualitatively different proteins are induced by low and high doses of radiation like those developed on the ground at various locations on our planet during the geomagnetic excursions. It also demonstrates, in zones with high background radiation, that high radiation would generate a beneficial radioprotective effect.

Studies of Brazil, China, India, and Iran have shown that radiation indirectly increases normal cell mutation. For example, natural radioactivity on human mitochondrial DNA was found to support an acceleration of the DNA mutation mechanism strongly.

Studies on the effects of chemotherapy on subjects with cancers indicate that, as a result of the radiation, the performance of the oculomotor system changed. The smooth pursuit and saccade onset times were shorter, and saccades had disproportionately shorter amplitudes. It means that a shorter saccade amplitude characterizes a reduced level of detail, which can be observed or revealed to the human mind.

P. Thomas Schoenemann from Indiana University, in his paper titled *Evolution of Brain and Language*, indicates some novel genetic information that contributed to the evolution of language. He explains:

Language evolution represents a special kind of adaptation, in part because language is a complex behavior (as opposed to a physical feature) but also because changes are adaptive only to the extent that they increase either one's understanding of others or one's understanding to others.

there is a complex interplay between language evolution and brain evolution. The existence of language presupposes a brain that allows it. Language must be learnable by the brains of children in each generation".

Circuits that result in a conceptual understanding that is relevant and useful to a given individual (ever-changing) environmental realities will be selected for and will spread over evolutionary time. This means that the adaptiveness (reproductive benefit) of an individual's particular behavior at any given moment in time depends crucially on the flexible responses of others in the group, who are at the same time attempting to behave in an adaptive manner in response.

The evolution of brain circuits, therefore, cannot be understood independent of the evolution of language, and vice versa. Language evolution is therefore inherently more complex than the typical evolutionary scenarios for physical characteristics.

The benefit of any particular mutation affecting linguistic ability is dependent on the existing cognitive abilities of others in one's social group. In it possible for mutations relevant to language to be adaptive strictly at the individual level (and therefore spread) even if they are not immediately useful for communication but if they are beneficial for some other reasons.

It might be that mutations responsible for neural circuits were initially selected for because of their usefulness for some nonlinguistic cognitive functions, perhaps by marking reasoning or thought more efficient or useful. In this case, however, these circuits would necessarily be nonlinguistic (and communicative), initially. Once they spread sufficiently throughout the population, the language could evolve to make use of them. Language adapted to preexisting brain circuits eventually prevails rather than causing the creation of wholly new language-specific circuitry.

Any changes in brain circuitry relevant to language in a given generation would likely consist of slight modifications of circuits that already exist, rather than major changes in the ways language is processed by the brain. This would be true for every generation, and the language, in the long run, would necessarily be continually biased toward the modification of preexisting mechanisms, rather than the accumulation of wholly new components. As a consequence, we should expect language circuits in modern humans to show extensive homologies with preexisting systems in closely related

> animals. Even if language evolved to use circuits not originally linguistic in function, these hijacked circuits would likely also represent modifications of nonhuman-specific circuitry.
>
> Language itself facilitates thinking and conceptual awareness. Increasingly complicated brains led to increasingly rich and varied thoughts, driving the evolution of increasingly complicated language, which itself facilitated even more complex conceptual worlds that these brains would then want to communicate.

On the other hand Hawks et al., 2007 states:

> Gene-culture coevolutionary models repeatedly demonstrated more-rapid responses to selection than conventional population-genetic models, which helps explain the argument that culture has accelerated human evolution. Conversely, under different circumstances, culture can also slow down genetic changes (Feldman and Laland, 1996).

Returning to my opinion about the role of cosmogenic radiative causes in shaping slightly, softly, gradually the brain architecture and its communication by language, I have to highlight the role of Solovki geomagnetic excursion that lasted from 7500 to 4500 years ago or for almost 3,000 years.

During epochs of high doses of radiation (caused by geomagnetic excursions, multiple generations inhabiting high background radiation zones, and long-term chemotherapy), the latency of the oculomotor system is significantly reduced. What does this mean?

LATENCY INTRODUCES PROBABILITIES AND OPTIONS TO THE MIND

Latency forces the brain to guess about a visual blind-spoti.e, to make predictions to compensate for this moment of blindness. Such a prediction covers the duration of this time gap or blind-spot, but this prediction is based on probability.

During a geomagnetic excursion, multiple generations living in that particular time would experience a significant reduction in the latency of their oculomotor systems. Those people would have a more accurate identification of the physical environment compared to people living during a time when these excursions are absent. But also they would encounter more mutations than normal.

How did this affect migrating humans? The decrease in latency allowed them to perceive better the physical elements of new environments where they were traveling, increasing their efficiency.

As another result of reduced latency, decision making became faster as their brain processing speed increased.

GEOMAGNETIC EXCURSION EFFECT ON LAGUAGE EVOLUTION

How did this reduction in latency affect the development of language?

Around forty-three thousand years ago, the language already existed among all humans but it was a virtual; static space of the mind; populated by nouns, pronouns, and adjectives. Because of its static nature, no significant language-related latency was generated.

Around twenty-five thousand years ago, verbs were added to the language. The reduction of visual latency for motions occurring in the environment being triggered by the radiative aspect of these excursions was like a pulsation of several radiative ranges with alternating patterns of reduction and expansion, which intermittently came and went away along the entire interval of sixteen thousand years.

Every reduction of visual latency was perceived as an increase in dynamics. The apparent increase in speed may have forced the brain to reflect such increased dynamics thus, the introduction of verbs and adverbs as virtual explicates of motion content generated by conversation was a natural result.

This novel introduction produced the verbal description of action; verbs inserted a pattern of virtual action within the virtual space of conversation that, until then, evolved as a static space of the mind with no virtual motion in it. This mental phenomenon, building up the new dynamic virtual space, mimicked the same mechanism of latency produced by the dynamics of motion in the visual analysis of scenes.

It is probably how language-related virtual action was created, and this action generated its latency, or I would say this is the virtual action latency of language.

It is safe to assume that the mentioned group of geomagnetic excursions had a significant impact on human genetics, also affecting the way language itself metamorphosed from virtual static into the virtual dynamic.

In short, in this time interval verbs and adverbs were introduced to the human language, creating virtual-latency of language, and the brain started to fill up this latency with goal-related materials.

Together, these factors generated what I call the Modern Human Brain, which became significantly different from the previous functioning of the brain.

From this point on, humans benefited from a gradual increase in goal-oriented activities, and where the goals were programmed to control, becoming technology.

As an immediate result, thousands of years later, roughly fifteen thousand years ago, the first attempts were made at inserted control, manifested as the domestication of dogs and plants among the Natufians of the Middle East and the people of the exposed shelf of the East China Sea.

VISUAL INPUT IS FILTERED BY THE OCULOMOTOR SYSTEM AND SLOWS DOWN THE ENTROPY OF THE BRAIN

Everything that decreases or obstructs visual input generates in response the increase in latency produced by the oculomotor system. Language has been a secondary task for the brain that acted similarly with an obstruction. However, in this category of obstructions also enter all mental illnesses, glaucoma, or simple effects like a decrease in luminance.

All such obstructions increase the size of the blind spot that is latency and consequently increase the guessing about the unseen space and its motions.

That is decreased resolution of the visual input increases latency.

The extent of uncertainty is measured by entropy: chaotic systems assume positive but finite values, random systems assume infinite paths, and the lack of uncertainty is represented by zero value. In the case of complex biological systems, their complexity grows over time generating a progressive ordering. The higher the complexity turns, the lower its entropy becomes.

The entropy of the visual stream results in about ten billion bits-per-second of raw data generated by the retina, of which only three million bits-per-second leave the retina for the brain's systems, with less than ten thousand bits-per-second coming under direct attentional scrutiny. This entropy decreases the signal more than one million times as it travels from the eye to the brain, meaning the tendency of this process of vision is to lower the processing entropy significantly in the human brain.

In an article by Paul J. Boon, Artem V. Belopolsky and Jan Theeuwes, published in PLOS, November 2016, entitled "The Role of Oculomotor System in Updating Visual-Spatial Working Memory across Saccades" the authors explain:

> "Visual-spatial working memory (VSWM) helps us maintain and manipulate visual information in the absence of sensory input. Taken together, results suggest that we neither rely exclusively on extra-retinal nor on retinal information in updating working memory representations across saccades."

> "Displacement of the saccade target caused a small but systematic bias in the direction of this displacement in all response modalities. This bias was smallest in the saccade task (21% of total target displacement), followed by the match to sample task (24%) and the mouse click task (36%)."

Here the average bias was 30 percent; this proportion indicates how much of our vision is "biased" or altered by the workings of the brain that comes in the form of extra-retinal signals. The authors found that:

> "Updating of remembered locations relies to a large extent on the extra-retinal signals."

> "There is a vast body of evidence showing that spatial memory and oculomotor preparation rely on the same neural circuits."

> "The relative contribution of retinal signals to the updating of a remembered stimulus seems to level off around 30%, independent of response modality."

Memory defines conscious processes. The 30 percent contribution of retinal information indicates that roughly 70 percent of generated memory is the contribution of something else that is not related to retinal information. As long as memory is what defines the conscious processes, our consciousness directly relies on vision for only one-third of its basic information; the neuronal circuits of the brain provide the remaining two-thirds. In other words, the rest is a fantasy and a hypothesis.

DISCREPANT MUTATION RATES OR CHANGES IN GENE EXPRESSION

These prehistoric people did not evolve as the rest of the populations placed mostly in Eurasia and North Africa. Why? In my opinion, the geomagnetic variation and its radiative effect was the primary cause of a mutagenic discrepancy. It produced a significant mutation in human brains, and the mutation opened a window of new opportunities for some people but no such window occurred for the others who were not exposed to the cosmogenic radiation.

Language, along with the incipient mechanism, played a role in communicating new unveiled options within communities making humans avoid a general path of extinction that manifested in Neanderthals, Denisovans, and some other hominids that we do not know much about but who probably existed before forty to ten thousand years ago. When the change toward new options emerged, it was associated with those vestibular mutations, associating language to technology development.

Geographically, there was a mutagenic discrepancy manifested between various locations and temporarily distinct associations of that mechanism to language itself. One area might become affected by one excursion or several of them, but another area may not be affected by any of them.

With possibly four to five excursions spanning a time interval of sixteen thousand years, one can imagine the extent of radiative stress caused each time at each location for such a considerable length of time.

One has to keep in mind this geographic differentiation that favored the case of oculomotor system single or multiple mutations in some people while not affecting any of those located outside the radiative impact area.

Dr. J. P. Lowery, O.D., M. Ed., with the University College of Optometry in Forest Grove, Oregon, explains: Nearsightedness and farsightedness have strong genetic components. If both parents have it, there is a good chance their child will be the same.

There are some significant environmental influences, such as near point work like reading. Other genetically linked eye problems include: color vision deficiency (color blindness) and Retinitis Pigmentosa (decline of the retina).

The story of current primitive people on our planet shows that people during different eras seem to have benefited from some novel options. Most of the cultures evolved based on those options up to some point, and if no other new options reached them, they would remain the type of development that is specific to a particular packet of options.

The Aboriginal peoples of Western Australia remained frozen-in-time as hunter-gatherers for almost forty thousand years after their arrival in Australia. New Guinea, Eastern Australia, and Amazonia's people remained frozen by their packet of options, which ended sometime in the Neolithic. Inca and Mezzo-America remained "frozen" in early Antiquity.

A different path was taken by social evolution in Eastern and Southern Asia. They evolved from the late Neolithic into a type of feudal farmer society that remained almost unchanged for thousands of years until two or three hundred years ago. They did not rely on slaves in their societies (the few slaves they owned would have only a temporary status) as was habitual in Western Asia, North Africa, and Europe.

A SHAPING BY ENTROPY

The most fundamental element that defines our brain workings is entropy.

In a paper, "Your Brain and the Second Law of Thermodynamics," the author, Dr. Hal McDonald, makes the following presentation:

> Dr. Robin Carhart-Harris has proposed a model of human consciousness in which two fundamentally different models of cognition are distinguishable by the degree of entropy which they exhibit, with "primary consciousness" characterized by a higher and "secondary consciousness" by a lower entropy level. Secondary consciousness is associated with the normal waking state in adults. In this state, the mind is organized in such a way as to represent the external world as precisely as possible, scanning the environment for predictable, stable patterns that presumably coincide with reality.
>
> The higher-entropy state of primary consciousness is less meticulous in its handling of the external world, representing reality in a manner that is easily biased by emotion.

To function effectively in the world, we depend upon a reliable mental representation of it, such as is provided by the lower entropy state of secondary consciousness.

In the realm of human consciousness, uncertainty has to do with patterns of perception, and how predictably these patterns cohere to the patterns that make up external reality. In the state of maximum cognitive entropy, we uncritically view all perceptions of the world as being equally valid representations of reality. During secondary consciousness, while the brain can enter a multiplicity of states and microstates, the system as a whole only ever entertains one winning state at a time. During the primary state, on the other hand, a reduced 'selectivity and conspicuity' extends the number of transient states that may be visited at any one time.

This high-entropy state of consciousness, characteristic of infancy, REM dreaming, and psychosis is obviously not well-suited to meet the cognitive demands placed upon us by our normal waking lives. A higher degree of cognitive certainty, or predictability, is required of us as we go about our normal day-to-day routines. Moving from the high to the low entropy end of the spectrum, uncertainty decreases and predictability increases until we reach a transition zone that is called 'criticality.' Just beyond the transition point lies a sweet spot of sorts where uncertainty is low enough, and our perceptions of the world are predictable enough, to allow us

> to go about our normal waking business. The sweet spot is limited, however, because too high a degree of predictability leads to rigid, inflexible thinking associated with repetitive and ruminative disorders such as depression, addictions, and OCD. At the other end of the spectrum, too much entropy is associated with a psychotic detachment from reality.

The role of secondary consciousness and its multiple states is to diminish uncertainty to a reasonable level and to assure a higher level of predictability.

Nabil Daddaoua, Manuel Lopes, and Jacqueline Gottlieb describe in their paper "Intrinsic motivated oculomotor exploration guided by uncertainty reduction and conditioned reinforcement in non-human primates," the way it works in monkeys.

> What are the cognitive and emotional factors that motivate animals to seek information? Monkeys are intrinsically motivated to search for and look at reward-predictive cues, and their intrinsic motivation is shaped by a desire to reduce uncertainty. The results suggest that free-viewing oculomotor behavior reveals cognitive-emotional factors underlying the curiosity-driven sampling information.
>
> The desire to predict the future is a powerful motivator in daily behavior and has been proposed to account for the extraordinary curiosity shown by humans and other intelligent species. Curiosity - the intrinsic motivation to learn - is arguably central to cognitive and social functions.

The initial interpretation of the findings with monkeys was that the animals placed intrinsic value on resolving "uncertainty" - that is, they sought to improve the quality of their beliefs about future outcomes. It was a preference for an early resolution of uncertainty.

In the present study, we addressed these questions using a novel paradigm where monkeys used rapid eye movements, or saccades, to search for or look at reward-predictive targets. The saccades were influenced by reward and uncertainty. The effects of reward were seen in the fact that the probability of making a saccade to the cue was higher if the cue signaled a positive rather than a negative outcome. The effects of uncertainty were seen in saccade timing. Longer fixation durations were not associated with positive outcome, showing that monkeys have a different reliance on overt versus covert attention. Some monkeys did not consider the saccade to be necessary for obtaining the reward. Monkeys were sensitive both to reward uncertainty and reward probability.

Our finding that a dual-component value function guides oculomotor behavior is consistent with more recent dual-process psychological theories suggesting that curiosity arises both from a desire to close "information gaps" (reduce uncertainty, or harvest information), and as a mere feeling of "interest" or "liking" of pleasurable items (conditioned reinforcement).

> Decision makers have intrinsic preferences for the early resolution of uncertainty and with accounts of curiosity as an attempt to fill an "information gap." It suggests that, even when subjects cannot increase their extrinsic gains, they value internal gains in information.

How is sensory information transformed into a motor response? A simple act of making a saccadic eye movement to a stimulus would take 60-150ms, but considering the amount of total transduction delay and neural transmission time, this sum may be 500ms.

Decision making requires choosing where and when to shift the eyes, but it also requires attention. Attention would be our ability to prioritize information. Eye movements (saccades) are the expression of this prioritizing pattern by the oculomotor system. This system allows us to see what the prioritization function looks like. Hence, attention is a prioritized state, and saccades are the motor capable of producing this state.

Humans tend to replace uncertainty with certainty which takes the form of options or probabilities.

This process is handled mainly by the oculomotor system and visual components of the brain that deal with latencies.

In the process of seeing, all mammals use a similar oculomotor system. The oculomotor system has to fixate on a visual target; when the target moves, between these moves, there is no reading of the dynamics, causing a blind spot. This move between fixations is called saccades, and it is the generator of the blind spot.

This blind spot is the source of uncertainty because we do not know what happened in this time interval or latency. The oculomotor system provides a solution: it compensates for the uncertainty of the delay with a prediction that is a probability, and it takes the form of correction saccade.

We know that, depending on the situation, the oculomotor system gives us the illusion of an increase or decrease in speed, and in other cases it makes us see more or less detail. Are these all pure and accurate corrections or are they, sometimes, a mash-up of various information?

Corrections and predictions contain additional information provided from the unconscious, but also an unseen reality.

Development of human language increased the latency, allowing the brain to increase the size of the probability. This augmented latency generated changes in the oculomotor system that forced it to process differently. This genetic upgrading spurred a new mechanism that provides, in part, a distinct type of information.

Augmented latency, together with the host of involved probabilities, manifest as options as they provide information that is regularly unseen and mostly unconscious. The same options are not generated for non-language-speaking animals.

THE EFFECT OF LANGUAGE ON OUR BRAIN

The human eye covers a visual field of about 200 degrees, while it receives detailed information from only 2 degrees that represent the fovea that is equipped with thirty thousand densely packed photoreceptors. During a time interval of 200-300ms, it is at rest and this is called latency or delay.

It takes about 80ms for visual information to reach the visual cortex. If the target moves, new information is used to trigger a corrective saccade. From the appearance of the target to the beginning of saccade in response to the motion of the target, a latency occurs. The first 20ms tend to remain the same disregarding the target parameters. The next 80ms or more are heavily influenced by the target type, speed, and acceleration. The first 100ms are an open loop with no visual feedback. This loop must be closed with information to produce continuous vision. For medium amplitude saccades, the latency is 200ms, but practically it varies from 100ms to 450ms.

It indicates that perceived visual reality is made of islands, and the saccades separate these islands. The oculomotor system builds bridges which connect the islands and give us the illusion of a continuum.

It is "strongly debated among scientist what kind of information is accumulated across saccades" that is the information making up the bridges. As it appears in several research papers, the "accumulated information is readily available after a saccade, biasing postsaccadic perception."

The research shows that "presaccadic information is used to produce postsaccadic perception." It is about "a transfer of visual features from a retinoptic into a spatiotopic representation." The experimental data seems to "support the hypothesis of a perceptual system where object representations can be updated spatiotopically across saccades, taking into account both object features and position." These quotes are from an article by Jasper H. Fabius, Alessio Fraccasso and Stefan Van der Stigche, published in Scientific Reports, 30 September 2016.

The oculomotor system will close the loop, providing (or not) information to fill the gap of 150ms–200ms. If no information is provided, then the latency will extend the processing 150ms–200ms. It means the processing time will be 150ms–200ms longer than real time. The entire mentioned process produces a decompression of the original information because the same information is displayed in a time frame that is now 150ms–200ms longer.

In this case, we would have a decompression of reality compared to normal visual processing. It is very tiny and occurs in all mammals (as shown, 3–4 times less in monkeys and seven times for dogs). For humans, 150ms–200ms remains unobservable and stands to a slower time-flow compared to true time.

During latency, the brain fills up this blind spot with some additional information.

What is this information? It could result from decompression of pre-saccadic information, and it can be supplied by the unconscious processes in the brain as a result of the slowdown of perceptual speed that produces a decompression of unconscious information.

Both processes occur in the oculomotor system: as the perceptual speed decreases the number of details of reality increases.

The studies on monkeys showed that their visual latency is smaller (40–70ms), compared to humans (150–200ms); the difference in the mentioned data is 110–130ms. Their apparent time flow gives the illusion that their time runs three to four times slower than for humans (and average monkey's life span is twenty to thirty years that in human years is sixty to eighty years of life; gorillas live forty to fifty years).

In the case of slower perceptual speed, the flow of time seems to speed up, increasing. The same process occurs during our lifetime: when we are young the time runs slow (because of higher processing speed), but when we are older the time runs faster (a lower processing speed). A similar process has occurred during human prehistory. In the Late Paleolithic and Mesolithic the time appeared to run slower, but going toward Antiquity, it appeared that the flow of time became faster. Today, we have the most rapid time flow ever experienced while our brain has the slowest perceptual speed yet encountered.

A HYPOTHESIS OF AUGMENTATION

Language influences the brain. This influence is described as the introduction of a second task—the task of language that overlaps on regular visual analyzing previously described.

LANGUAGE AS A SECOND TASK

For this exemplification, I have chosen a paper of Marcel Adam Just, Timothy A. Keller, and Jacquelyn Cynkar of Carnegie Mellon University, *A Decrease in Brain Activation Associated with Driving when Listening to Someone Speak*, published in 2008 by Elsevier. The authors state:

> Each component task-evoked much less cortical activity when it was performed concurrently with another task than when performed alone, even though the two tasks drew on different regions of the brain. Our research investigated the effects of performing an auditory language comprehension task while simultaneously performing a simulated driving task. Our sentence listening task did not require the participants to speak, and thus is less disruptive to driving than a full-fledged conversation might be. The tasks involving speech production did affect both eye-movement (saccades) and behavioral performance. Also, a verb generation task did cause disruption.

> The results are consistent with the hypothesis, derived from previous behavioral studies, that a simulated cellular telephone conversation disrupts the driving performance by the diversion of attention as reflecting a capacity limit on the amount of attention or resources that can be distributed across the two tasks. This capacity limit might be thought of as a biological constraint that limits the amount of systemic neural activity that can be distributed across parts of the cortex.

LANGUAGE AND ATTENTION

Another article titled *The Slow-Down Course of Visual Attention* published by Ward R., Duncan J. and Shapiro K. in *Cognitive Psychology*, February 1996, indicates:

> Human attention depends on the number of attended objects, not on the number or complexity of responses. Visual attention is not a high-switching mechanism, but instead a sustained state.

This article reveals the effects on visual attention by simply listening and having a conversation. The first generation of human speakers has experienced such effects, and such effects probably and gradually introduced some continuously increasing forms of attention deficit.

This case eventually drove some generations into an increasing crisis where the environment turned riskier and more dangerous and where they needed new subsistence capabilities, like domestication of animals and plants, as well as shelter.

However, speaking and listening have an even more significant effect on eye movements that relate to the oculomotor system and its latency.

LANGUAGE AS A DISTRACTOR

Jonathan van Leeuwen and Artem Belopolski, in a paper titled *Distractor Displacements during Saccades are Reflected in the Time-Course of Saccade Curvature*, which was published in *Scientific Reports* (article number 2469 of 2018) state:

> Observer tends to miss object displacements if they occur during saccades." Here is a mechanism of saccadic suppression of displacement. Let's see how it works.
>
> It appears as if the oculomotor system contains accurate information about space, but that information does not necessarily reach conscious awareness. The predictions made by the oculomotor system before the saccade can be updated after a saccade, or after 190ms. It takes the mentioned system about 250ms to update from old information to new information.
>
> It is widely accepted by the literature that visual changes that occur during saccade are hard to detect and often escape conscious perception. The reason is that the conscious tends to preserve the image of world stability. Thus, our conscious perception benefits from perceiving the world as stable. But the same mechanism enables us to unconsciously adapt our actions to changes, which we do not consciously perceive.

> The present results further demonstrate that this prediction drives the oculomotor system up to approximately 190ms after the saccade.

The change in the details of the dynamic environment which occur during the saccades are recorded only by the unconscious processes, and they are provided an unconscious type of guidance to the conscious state.

The brain connects the information that occurred before the saccades with the information obtained after the saccade — omitting the information change that happens during the saccade that is the latency interval. As it seems, latency is the reason the speed of perceptual processing slows down. Latency is estimated to be 190ms. The authors indicate that the lack of information for the time duration of the saccades is replaced by a prediction of what might happen during that particular time.

LANGUAGE AS A FILTER OF REALITY

J. M. Delgado-Garcia, E. Godaux, and P. P. Vidal in a paper, *Information Processing Underlying Gaze Control,* published in *Elsevier Science* (2013) explain:

> Clearly, the smooth pursuit system is an attentional focusing mechanism that spatially filters visual motion inputs, and if necessary, overrides the optokinetic response.
>
> The oculomotor system deals with generating compensatory eye movements.
>
> The delayed component evolved as a back-up that assures the decomposition of the optic flow pattern.

It appears latency participates, as a vestibular mechanism, in the decomposition of the optic flow.

This 190ms latency may request feedback from the brain network. What is the content of this feedback and where does it come from?

Does a decompression of unconscious information intervene here?

If so, then the feedback is a decompressed unconscious information that regularly is not available for conscious processes. Here, the addition of information from the unconscious fills up the space-time of a gap that became too large to be handled by normal constraints of the brain network.

> The new vestibular mechanism was selectively sensitive to linear accelerations, and its output scaled inversely with viewing distance. The smooth pursuit subsystem spatially filters the visual motion inputs, utilizing an attentional focusing mechanism to specify the location of interest.
>
> Speech input modifies attention states, leading to a bias towards one object, as speech directly describes or indirectly refers to.

The authors of this paper refer to the filtering process that acts as a compression and decompression of information while the compressed information results as a type of order where the variables are manageable for the linear working mind. They also explain that new vestibular mechanism favors language that changes the brain workings by modifying the attention states.

A REVIEW

These two papers mentioned above provide some fundamental information for the journey of correlations I intend to do in this research of mine.

To summarize, language is a second task. Language is a distractor. Language involves attention states that it modifies. Oculomotor saccades produce a lap of information to conscious activity with a 190ms latency. Language reflects the existence of a vestibular mechanism responsible for latency. Language slows down the perceptual speed of the brain by 190ms. The latency component helps to decompose the optic flow in sub components. This vestibular mechanism acts as a filtering system that generates a linear world of order by generating a continuum. In the end, the vestibular mechanism favors the attention focusing on near field objects, while it tends to ignore the far-field ones.

These five or so fundamental processes take place in the brain of all mammals. In monkeys, the latency is 120ms shorter than in humans depending on what research data we quote. The known latency for monkeys amounts for 70ms (40ms from other sources). While the research mentions 190ms in humans, the 120ms seems to be the distance traveled during our evolution until the revolution in language occurred some forty to thirty thousand years ago.

If the distance in latency between humans and monkeys is 120ms, then one can say that this mechanism at the time of the mentioned language revolution was already significantly more evolved in humans compared to other animals.

Because of a bigger latency, humans became more visually distracted than the monkey. The human gain in language generated a decrease in their environmental skills. But this was partially offset by progress in the design and construction of tools.

Is this high latency in the human oculomotor system what favored the next step of producing a better language forty to thirty thousand years ago? Was there a cosmogenic cause being generated in that time interval by eight thousand years of excursional mode, that was developing only in the Earth's environments placed north of the 45 degrees parallel, which exists only out of Africa? Or, is the Southern peak of South Africa placed on a 35-degree parallel still a match for the southern geomagnetic flips?

What caused high oculomotor latency in humans in the first place? Was the Gaotai geomagnetic excursion, with the length of almost ten thousand years (which occurred eighty-three to seventy-three thousand years ago and which also reached the African continent all the way to its southern peak), the cause placed within Africa that suddenly produced the first increase in latency around eighty-three to seventy-three thousand years ago? Did something similar occur before that time?

AN ANATOMICAL TECHNOLOGY

There have been some special conditions—environmental and radiative, which have favored a special genetic adaptation that found its niche in a new system of communication, and this new system evolved as the human language.

An anatomical technology that was only possible and available for humans was needed at that particular time, and it helped humans survive those new critical conditions.

This vestibular mechanism was easily changed to accommodate language. During this change, latency was significantly increased and the latency/delay then further slowed down perceptual processing. As a result, an increase occurred in the processes of decomposition and decompression which in turn provided, in one way or another, new details to the human mind.

The gradual increase in the richness of details produced options exploited by humans to build their civilization.

None of those options became available for non human animals because they remained deeply hidden in their unconscious.

LANGUAGE AND BRAIN ARCHITECTURE

In a study published by Gerry T. Altman in June 2011 in *ACTA Psychologica* (Elsevier), the author analyzes several studies and experiments dedicated to the perceptual speed of processing in human subjects.

> How long does it take, as the language unfolds, for the oculomotor system to demonstrate sensibility to the distinction between "signals" (eye movement due to the unfolding language) and "noise" (eye movement due to extraneous factors)?
>
> The earliest discrimination between signal and noise occurred around 100ms.

He found that the average visual perception of a scene takes from 100 to 200ms. He also found that, while speaking, listening is involved. It supplies an additional latency in processing that amounts on average to 150ms.

The conclusion would be that the introduction of language has produced a latency in perceptual processing that caused the process to extend from normal average 150ms–190ms to 300–340ms because of the new 150ms latency.

Mishra R.K., Oliver C.N., and Huettig F. found (Elsevier, 2013) that "language mediated eye movements appear to be for the most part unconscious and have all the hallmarks of an overlearned behavior."

Willems R.M., Oostenveld R., and Hagoort P. (Elsevier June 11, 2008) concluded their study with:

> We conclude that although full semantic integration of linguistic and visual information occurs through a common mechanism, early differences in oscillations (of alpha and gamma band power) in specific frequency bands reflect the format of the incoming information and, importantly, an early context-based detection of its congruity with respect to the preceding language context.

It implies that at the beginning of language, there were distinctions between the brain rhythms for vision only and language only.

> Stephen Levinson of Max Plank Institute for Psycholinguistics explains (2015):
>
> The brevity of these silences is doubly astonishing when you consider that it takes at least 600 milliseconds (MS) for us to retrieve a single word from memory and get ready actually to say it. For a short clause, that processing time rises to 1,500 milliseconds. It means that we have to start planning our responses in the middle of a partner's turn, using everything from grammatical clues to changes in pitch. We continuously predict what the rest of a sentence will contain, while simultaneously building our hypothetical rejoinder, all using largely overlapping neural circuits.

LANGUAGE ARISES FROM A NEAR-FIELD OCULOMOTOR ADAPTATION

Another study titled *Simple Reaction Times to Cyclopean Stimuli Reveal that the Binocular System is Tuned to React Faster to Near than Far Objects*, published in *PLOS One* on January 5, 2018, shows that

> "...convergent evidence from psychological and neuropsychological studies have revealed asymmetries between the processing of near (crossed) and far (uncrossed) binocular disparities. The results showed that disparity and contrast had significant effects on reaction times (RTs).

Sensitivity of the human visual system is biased for near versus far disparities, and near stimuli can result in faster motion responses, probably because they bear higher biological relevance.

This study indicates that the human eye favors near field objects as a matter of ecological adaptation. The resolution on such objects increases. Larger objects are favored versus smaller ones because they potentially represent more dangerous or more rewarding animals.

This environmental adaptation for a better description of near-field objects became significant when spoken language developed throughout human groups. The language used the door opened by this adaptation to the natural near field approach. It favored the conversation between individuals located closer to one another. Face mimicry, eye gazing, and gestures also played a role in prehistoric conversations.

The time referred above is computed from the onset of the word by a speaking person to the moment when the oculomotor system of another person determines the appropriate scene's target to fixate.

During the experiment, the delay appears between the speeds of the perceptual processing in the subjects of a speaker-listener team compared with the visual processing speed of individuals who are not involved in speaking-listening but merely watch a scene. Their processing speed was 150ms. That was a time equal to the average delay of 150ms encountered for the speaker-listener groups.

It means that a speaker-listener will process in, at least, double-time compared with a regular visual but silent observer.

Many other papers debated this same subject, which was the result of multiple experimentations and interpretations.

The latest results based on MIT research reach a significantly different conclusion, indicating that the fastest speed for processing a visual scene by a human is 13ms. The saccadic delay remains the same at 150ms.

The language revolutions, naming here only those placed seventy thousand years ago and forty to thirty thousand years ago, were responsible for changing the speed of human perceptual processing. In this case, by a decrease of more than eleven times, considering only the MIT study.

A study by Marc J-M Mace, Guislaine Richard, Arnaud Delorme, and Michele Fabre-Thorpe with the Faculty of Medicine, Toulousein, France (December 2004), found that three monkeys performed a categorization task with a delay of 20ms compared to a 40ms delay previously found for humans performing similar tasks. Monkeys always produced their motor response faster than humans. Express saccades are seen at 70ms in monkeys and 100ms in humans. As it appears, saccades in the monkey occur in less time, transferring information within any cortical area. It means that monkeys do not have to predict to the same extent that humans do, while their visual interpretation is more accurate than it is for humans.

Modern humans, in the last thirty thousand years, may have lost at least 1/3 of their vision acuity. If this is correct, then the differentiation between processing speeds of vision and language would increase further to 15 times or more.

In general, it became obvious that the spoken language has influenced the perceptual processing speed of the human brain in the sense of slowing it down.

ENTROPY SHAPES FROM SKULL SIZE TO THINKING

A paper by Maciej Henneberg titled *Decrease of Human Skull Size in the Holocene* published in *Human Biology* in June 1988 states:

> Craniometric data collected from approximately 3,300 individuals, dated from the Upper Paleolithic to modern times, were used to evaluate the trend" observed in human skull changes.
>
> Among male samples, the peak CC (Cranial Capacity) occurred in Mesolithic (1593cc); the lowest value falls in modern times (1436cc). In female samples, the timing is the same.
>
> A decrease of 157cc (almost 10%) in male and 261cc (almost 18%) in the female is a considerable one, of an order of magnitude comparable to the difference between averages for H. erectus and H. sapiens.
>
> But this may not be a result of the direct operation of selection upon brain morphology.

It implies that something else, like language development, was one of the real causes for changes in the size of the skull.

The diminishing in size of the human skull reflects the gradual evolution of human language. In the epoch placed seventy thousand years ago, incipient language development was minimal and stayed that way until thirty thousand years ago. The last language revolution probably caused a gradual decrease in skull size. The peak decrease occurred during the Holocene Era that started ten thousand years ago. Most scientists agree that gracilization of the human body has been an important factor in the diminishing of skull size. No attempts have been made to collaborate the impact of language development with the concurrently observed reduction in skull size.

Both gracilization and language development contributed to the reduction in skull size, but something else more important intervenes.

In a paper published in Physical Review Letters, on July 11, 2016, one of the authors, Dr. Sebastian Goldt from University of Stuttgart, Germany, explains that his team has discovered a very important phenomenon. "The slower a neuron learns, the less heat and entropy it produces, which increases its efficiency."

It shows that the introduction of latency in the human brain due to development of language has directly contributed to a significant decrease of perceptual speed; this decrease in perceptual speed has generated a decrease in neuronal processing speed. Humans begin to lose intelligence. But the new study from Germany has discovered the contrary: the decrease in neuronal activity reduces entropy, increasing efficiency. However, this reduction in entropy helps reduce the brain metabolism so that it does not need to produce as much energy as it did before the increase in latency. Potentially, this would be the best candidate for a cause of the reduction in the size of the human skull.

The evidence about the beginning of the process of reduction in human skull size is well represented in the archaeological records circa twenty thousand years ago. This date also validates the presence of new patterns in neuronal workings.

It also shows how neuronal mechanism patterns developed rapidly through the Holocene Era.

Visual images are great for receiving knowledge but are far less adequate for expressing knowledge and communicating. There was a tremendous need for a better way to communicate. An adaptation imposed it to new environments.

Nicolas Gompel and Benjamin Prud'homme in the paper, *The Cause of Repeated Genetic Evolution,* published in *Developmental Biology* on August 1, 2009, explain:

> The repeated evolution of similar traits in organisms facing similar environmental conditions is a pervasive phenomenon, including for animal morphology, and is considered strong evidence for adaptive evolution. Case studies reveal that although multiple genetic paths were often possible to evolve a morphological trait, similar evolutionary trajectories have been followed repeatedly in independent lineages, suggesting that biases influence the course of genetic evolution.

LANGUAGE PROGRAMS THE MIND

Peter Carruthers said, "Conscious thinking is conducted by means of natural language sentences." As it appears, they evolved in connection to one another.

Another concept expressed in the paper, *How Language Programs the Mind,* published (2016) by Gary Lupyan of the University of Wisconsin, Madison, and Benjamin Bergen of University of California at San Diego.

In their paper, the authors conclude:

> We have stretched out a view on which language is a critical tool for programming the mind. The idea is intellectually aligned with Clark's view of language as a "computational transformer which allows pattern-completing brains to tackle otherwise intractable classes of cognitive problems. (Clark, 1998, p.163) and resonates with Dennett's idea of language as a "cognitive autostimulator" (Dennett, 1992, see also 1996).
>
> Language, in this view, is not simply a communication system; it is a control system.

Here, control comes from programming.

When people talk, they program each other, trying to extend their control onto each other.

The question is: What allowed the mind to become programmable? One more idea is that "language directly interfaces with mental representations, helping to form compositional, abstract representations, which are not existing apriori."

The fact that the human mind became programmable (some time ago, perhaps forty thousand years) is simply witnessed by the systems of teaching we employ today. Through teaching, we program others to learn what we want to be learned. After learning, people, like software, stay instructed until something erases it.

The research of the German team from Stuttgart indicated that "the slower a neuron learns, the less heat and entropy it produces which increases its efficiency." Slow learning appears to be the change that made the human brain programmable versus other animal brains which remained unprogrammable or only capable of processing very little programming.

Time and again, changes in human brain architecture that occurred some thirty thousand years ago affected the way the human brain worked. A single new mechanism became responsible for all sudden changes in prehistoric humans, favoring the "latency" that slowed down the entirety of neuronal activity. Once the new mechanism became fully functional, humans turned into a very different animal, distinct from non speaking animals.

THE BRAIN SELECTS NEW MODALITIES OF WORK

In the paper, *Execution of Saccadic Eye Movements Affects Speed Perception,* published in PNAS on February 27of 2018, the authors conclude:

> We investigated the effect of corrective saccades on speed perception...and found that the direction of corrective saccades influenced whether the observer perceived the tracked moving targets to be moving faster or more slowly than in pure pursuit trials. We also found a correlation between the average eye velocity during each trial and the perceptual judgment.

Another PLOS paper titled *A Hierarchy of Time-scales and the Brain* published on November 14, 2008 in *Computational Biology* concluded that:

> ...neural rhythms in the brain play a role in speech, speech segmentation and coding of speech at multiple time-scales. These models (as brain computation models) reduce high-dimensional input to a few variables or 'causes' in the environment. These environmental 'causes' do not need to be physical objects but can be any quantity that predicts sensory input. A conflict

> arises between different and similar time-scales, like between visual and auditory streams.
>
> Fast dynamics can be predicted by supraordinate representations that change more slowly." It shows how our world of "order" is generated by a selection of variables that change into more slow dynamics.
>
> We have proposed that the brain employs a hierarchical model. At lower levels of this hierarchy, close to primary sensory areas, neuronal states represent the trajectories of short-lived environmental causes. Critically, at each level, representations depend on and interact with, representations of other levels. Cortical structure recapitulates a hierarchy of temporal scales.

Brain Rhythms of Language: Nouns Versus Verbs published by *The European Journal of Neuroscience* in May 1996 concludes that

> these results suggest that local high-frequency brain responses can be indicators of conscious processing of motor and visual associations of verbal material. It provides additional evidence that nouns and verbs have distinct neuronal generators in the brain.

It shows how the architecture of the brain changed because of the evolution of language. If verbs and adverbs were introduced in language forty to thirty thousand years ago at that time the aforementioned neural generators appeared in the human brain. Was there a neuronal mutation?

> Predictions about sensory input at fast time-scales become imprecise when projected too far into the future. One way to deal with this uncertainty is to use concepts to guide representations at shorter time-scales.

Another anatomical change occurred in the brain when it began to select shorter time scales.

STATISTICAL ACCUMULATIONS PROPEL LANGUAGE JUMPS

Research on childhood development shows that the development of language seems to occur in children once a statistical accumulation of data reaches a sensible threshold. At this threshold, children suddenly exhibit important qualitative processes in acquiring language.

Perhaps a statistical threshold similarly built up in the language and art explosion manifested thirty thousand years ago. This explosion was an expression of a new function in language development. The new function may be the expression of a new quality, and it may be a mathematical rule that governs the transitions from one quality to another. Did the brain's statistical accumulation favor a mutation?

It also provides a picture of the development of language where all its branches encounter unconnected qualitative jumps.

BACK TO THE ORIGINS

In a paper by Hat Whitehead (of Halifax University, Canada) and Peter J. Richerson (of University of California at Davis) titled *The Evolution of Conformist Learning Can Cause Populace Collapse in Realistic Variable Environments* published by Elsevier in 2009, the authors explain:

> If the environment of an organism varies considerably, then, the direct genetic determination will not allow the behavior to track the environment efficiently. Once a species is using individual learning to track environmental variation, then a less- costly strategy of social learning may invade.

Was this the case in humans migrating out of Africa? They needed environmental efficiency, as provided by a new type of communication that took the form of an evolved language.

Further on, the authors provide evidence that "the development of language generated social learners, who use information provided by the behaviors of others to regulate their behavior, through mechanisms such as imitation and emulation."

The Out-of-Africa migration experienced an explosion in the quality and capability of language (forty to thirty thousand years ago) that also produced social learners who were balanced by progressive individual learning about the environment. The migration itself introduced individual learning on a larger scale. Soon after this migration occurred, the language revolution initiated a significant increase in the size of populations and the development of social learning that was intermediated by the creation of better language.

The authors continue:

> Social learning can lead to the homogenization of behaviors among individuals in a group and among groups. By some definition this is culture.
>
> It is clear that social learning and culture were not possible without language development.

> However, as this study shows, in our modern era "the behavior of the population will no longer be able to track environmental variations, as the individuals will only be reacting to each other and not to the underlying environment changes."

It is also possible that a similar situation has been reached many times in our prehistory and history, generating crises and social collapses. Such a situation creates conformism that prevents any adaptive responses.

The migration out of Africa, estimated to have occurred between seventy to sixty thousand years ago, was generated by a sudden cooling seventy thousand years ago; the glaciation in the mountains of Morocco, Algeria, Kenya, and Ethiopia; an extension of Sahara and Kalahari deserts; a multifold increase in atmospheric dust; and a general drying out of the climate.

As a result of glaciation, the sea level dropped considerably up to 135m causing several sea straits to turn into shallow passages, making them easier to cross.

It is estimated that the African population, seventy thousand years ago, was around thirty-five thousand people. According to Dr. R.I.M. Dunbar, the average size of prehistoric human groups was 150 individuals. Before the migration out of Africa, nearly 250 human groups existed. Genetic research indicates that the people who left Africa through the Sinai Peninsula and the Bab-el-Mandeb Strait were only two thousand individuals.

However, it is possible that there were several waves of migration separated by brief intervals of time. Groups probably crossed into Sinai, and at the same time, other groups crossed the Red Sea from Ethiopia/Somalia to Yemen.

It is also possible that during this climate crisis, other crises mounted and they were social. These resulted from a weak and inappropriate response to environmental challenges.

DUNBAR'S HYPOTHESIS ON ARISING OF HUMAN LANGUAGE

Dr. R.I.M. Dunbar, a professor of anthropology at University College of London in his famous paper, *Co-Evolution of Cortex Size, Grouping Size and Language in Humans* (1993) also makes several suggestions about the evolution of language in humans. Dunbar found that Paleolithic humans socially groomed and partook in non-foraging activities some 42 percent of the time, while chimpanzees spent some 25–43 percent of their time grooming. If sufficient time spent in grooming cannot be maintained, then the group tends to fragment.

Dr. Dunbar considered that language evolved as a cheap form of social grooming because it allowed one individual to do two different things at once.

He concludes his paper by saying:

> This would seem to suggest that many of the basic properties of speech and language were already available in the more advanced nonhuman primates. What was required was their close integration and elaboration, and this may have been dependent on a significant increase in neocortex size to provide the necessary computing capacity.

LANGUAGE DEVELOPED THE ABILITY TO MANEUVER THE CONTENT AS RECURSION

Most theories still support the idea that human language originated in the brain capacity for recursion that shows the ability to make chain combinations, as in AABB and ABAB and so on. Hauser, Chomsky, and Fitch (2002-2005) proposed the theory of recursion as the origin of language and its grammar.

Language development reflects a linear computational ability which is not unique to humans. All organisms tend to make estimates over content that are balanced with another ability to estimate context. The understanding of context was a primordial capacity that assured the survival of a species.

Here, recursion is the ability to maneuver content and this human ability side-tracked into aural maneuvers. It helped select aural harmonics and arrange them into particular sound expressions. Not coincidentally, human language is tonal and its development went hand-in-hand with music and art which both display a mix of harmonics.

The development of language was the sole contributor to a considerable rise of linearity as recursion in human thinking.

LANGUAGE RECRUITS NON-LINGUISTIC NEURONAL CIRCUITS

Dr. Mark H. Bichard shows experimental proof that "pre-existing psychological, perceptual and motor capacities would have been recruited for the task of language learning and use. These capacities would have been honed and specialized further by natural selection ... specialized for language-related tasks [such] that they cease to perform any non-linguistic functions at all." When this happened, the volume of the brain itself had been significantly reduced.

He also said, "Human beings perceive and understand the world through the senses, and then, that epistemic connection with the world occurs via transmission of information from the world through those senses to the mind. The converse perspective, on the same assumption, is that the environment influences individuals, both microgenetically and developmentally, via the information that is generated in that environment and transmitted into the minds of those individuals. However, people select what they value. The idea of cumulative niche construction of values has obvious applications to the case of language."

In general, one can say that language has jumped from a finite communication system with many words, the incipient stage of language development, to infinite recursive language where it later embedded its acquisition of mental analysis and synthesis which manifest in the forms of math and logic.

One can assume that the explosion mentioned above, probably thirty thousand years ago, produced verbs and adverbs that suddenly transformed the language into a rich and complex means of communication.

PROCESSING SPEED OF PERCEPTION

Processing speed is the pace at which you take information, make sense of it, and begin to respond. The information can be visual or auditory, such as spoken language. It is possible that the brain's processing is limited to certain quantities of details and when the number of details increases, the brain has to slow down processing to fit its quantitative capacity of operation. In reverse, when the number of details diminishes, the speed of processing increases.

In a paper published recently by Dr. Cherry Florance, the author expressed the idea that the...

> dominant visual brain interferes with the activity and development of the auditory-verbal brain. The visual brain uses opticoders, which decode the world around them through visual analogies rather than verbal reasoning. The lexicoders decode the auditory signals, helping to express knowledge, and helping with communicating those expressions (of knowledge).
>
> The opticoders brain is considered 20,000 times (60,000 times from other sources) faster than the lexicoders brain, leading to a type of outside-the-box-thinking that resulted in some of the greatest inventions and creative contributions.

By slowing down their perceptual speed processing by at least twenty thousand times, brains with evolving lexicoders reveal an ever-increasing quantity of details. The gradual advance in the decrease of speed produced a gradual increase in details.

> Our education system devotes the majority of its time training us to improve spelling, writing, reading, and speaking in front of a group and comprehending reading passages, all of which require us to practice using our lexicoders in a step-by-step sequential building pattern.

The "Maverick Mind children think 20,000-60,000 times faster than their mouths can go." Is this somehow similar to a Middle Paleolithic mind and behavior?

> Consequently, their verbal communication suffers. Mavericks are unable to understand, think, and speak in a language familiar to verbal people. Their brains literally speed past the language in such a way that language does not properly develop.
>
> Mavericks may focus intensely on visual tasks to the point where they do not hear anything that is said to them and become very frustrated when interrupted.

Is this an artifact from the non speaking or rudimentarily speaking eras?

There is evidence to show that Dr. Hans Jenny, called the father of Cymatics, in his experiments on the acoustic properties of fluids, performed from1960 to 1975, revealed that when he increased the amplitude of the sound pattern, the motions became more rapid and turbulent. When he increased the frequency of the sound pattern, the number of details greatly increased along with an increase in complexity.

Some physicists and researchers of the same epoch did experiment on the saccadic movement of the human eye. They found increased saccadic amplitude corresponds to an increase in the perceptual processing speed and a decrease in the frequency or density of the elements perceived by the subjects. At decreased amplitude, significantly increasing density, resolution, or frequency of details was found.

However, there is proof that there is no physical correlation between amplitude and frequency in wave phenomena.

ADDING READING/WRITING TO VERBAL LANGUAGE

In a paper titled *Visual Processing during Natural Reading* written by Balazs Knakker and Zoltan Vidnyanszky and published on May 27, 2016, the authors explain several experiments they conducted to define processing speed during natural reading.

The experiments produced various distances between the letters of written texts while subjects were observed reading them.

The authors explain their conclusion:

> Significant main effects of letter-spacing were obtained for both fixation duration and saccade amplitude in eye movement measures. Fixation duration was reduced with the increase of letter spacing, suggesting that visual processing load is decreased as a result of increased inter-letter spacing. On the other hand, saccade amplitudes were increased with larger letter-spacing, which is in agreement with previous results."

One can conclude on the reverse process: when the spacing between letters is deceased, saccade amplitude would be lower.

> Reading was fastest at the smallest inter-letter distance and decreased with increased

letter-spacing. These results reflect those neural processes of visual information processing during natural reading.

Now, we can compare the data provided by many experiments about processing a visual scene (13–20ms) and about a conversation between people in which a listener can process 120–200 words/ minute. It is equivalent to dividing the result of the transformation of milliseconds into minutes. The result is six thousand; this means that the development of human language has produced in our brain functioning a diminishing in perceptual processing of six thousand times.

Here, one can conclude from the Hungarian research, presented in previous paragraphs, the direct impact of language on the human brain. The Hungarian authors found that when the saccadic-amplitude-processing-speed increases, the brain can distinguish fewer details from the reality (that reality was a printed text). Conversely, when the saccadic amplitude processing speed diminishes, the number of details increases, inversely proportionate to the density of the printed text.

One can say that the gradual evolution of language increased the number of observable details of reality. The increase was gradual. Thirty thousand years ago, when the qualitative revolution in language occurred, we may consider a smaller amount at the beginning of this process that gradually rose to six thousand times for a fully developed language (like the one that occurred about twenty thousand years ago).

Undoubtedly, an important revolution in language occurred at the time the writing was discovered which occurred in the seventh century BCE

The production of writing was well advanced by the use of symbols: the first symbols appeared in the cave paintings from Indonesia, France, and Spain circa thirty-nine thousand years ago. A stone tablet with symbols from China was dated twenty thousand years old. The Balkan tablets were dated 5500 BCE. The Sumerian tablets with cuneiform inscriptions were dated 3500 BCE. It was a gradual but long transition from spoken language to written form.

During this gradual transition, other changes occurred, like domestication of animals and plants, settlements, first villages and cities, horse domestication, the use of animal power, the use of new means of transportation, and the discovery of metallurgy.

Another aspect of language evolution was the perception that the flow of time sped up. This illusion would represent functional changes which occurred in architecture and brain processing, while they generated the illusion of a more dynamic life.

Thirty thousand years ago, the perceptual speeding of time developed gradually with extreme figures that we have experienced in the last two hundred years but significantly more abruptly in the last ten to twenty years.

Until two hundred years ago, human development does not indicate any exponential growth; the exponential course started only two hundred years ago.

All these changes reflect the options revealed to humans by the gradual development of language because language slowed down our perceptual processing and in response to this decrease in speed, a gradual increase in the available options appeared.

MODERN BRAIN CHANGES THE TIME AND SIZE OF PERCEPTION, INTRODUCING NOVEL OPTIONS

A recent paper, published in the journal, *Animal Behavior,* by a team of researchers from Trinity College Dublin, found that time perception varies among species and even among individuals of the same species. The leader of the team, Dr. Jackson, describes their results:

A whole world of detail is out there that only some animals can perceive, and it's fascinating to think of how they might perceive the world differently to us.

> They demonstrated their findings with a phenomenon called "critical flicker fusion frequency," which is based on the maximum speed of flashing lights an animal can see before it just looks like a steady, constant light - the same principle behind the illusion of non-flickering television.

One of the conclusions of this study is that smaller animals experienced a decrease in the speed of their eye perception or of the saccadic movement, similar in many respects to human speakers and listeners that allow them to distinguish more details of reality. It was found to be an adaptation for survival. In contrast, larger animals process perceptual information, and thus, they see significantly fewer details as a result.

We know that in the last 130,000 years, 177 species of animals larger than 10 kg became extinct. Many became extinct in the last 30,000 years in areas where humans were in extremely few numbers, like in the Americas and Australia, and humans were not the cause of these extinctions.

The early stages of the human language, which can be dated around 130,000 years ago, coincided with the beginning of this mass extinction. Did the introduction of human language coincide with an ecological turn in animal evolution toward a type of smaller animals? Were humans adapting to this planetary trend by reproducing in themselves those characteristics of small animals? We can see today that language makes us behave like smaller animals.

So, was language that part of needed adaptation? It seems that it was. Is the enlarging of the saccadic latency a reason for an apparent longer lifespan in humans?

The revolution in writing and reading also developed gradually because, to influence further brain architecture, it was necessary to develop a mass phenomenon. However, in 1975, only 51 percent of the world population was literate, and only after the year 2000 can we discuss a significant mass phenomenon.

At present, some 750 million people are illiterate, or 10 percent of the world population, and also 10 percent of the young generation falls in this category.

New research and experiments, comparing the visual perception of a scene with writing/reading of a text, have found the difference to be now sixty thousand times.

So, from spoken language development to writing and reading, the decrease in the speed of perceptual processing increased another ten times, reaching a total of sixty thousand times difference in processing speed.

FAST READING

One of the most eloquent examples is the scientific research on fast reading.

Here, when the speed of reading increases, it was observed that the reader could not recall any of the details described in that text he/she was reading. In this case, the increase of perceptual speed produces a clear diminishing in the observed number of details.

Fast reading research discloses a lot of things of significant importance.

The fast reader can understand the text in general, and that understanding defines the context and its options.

The second thing is that the fast reader cannot recall the details described in the text. Here occurs a very poor content disclosure.

The experiments analyzed the behavior of the subjects' brains at various speeds of reading. At maximum reading speed, the details of the text were missed but the context was not altered. As the speed decreased, more details became observed by the subjects.

This experimental situation can be compared to the impact of language evolution on human populations. Before the language revolution, humans used a type of language that made them observe very few details of the environment but gave them a strong understanding of the context. As language evolved, the details of the environment increased, revealing more content, but the understanding of context meaning began to diminish gradually.

Today, content grows exponentially; doubling every year, but we are losing significant meanings from the context.

Other experiments concluded that, on average, a good reader could process 250- 300 words per minute and a listener can process 120-200 words per minute. It reflects the progress made by writing/ reading compared to a spoken language, as a faster means of communicating.

THE AGE OFEXPONENTIAL GROWTH

In much of the time interval from thirty thousand years ago to the eighteenth century, the estimated advances in human society showed an increase in the range of 0.01 percent - 0.10 percent.

At the end of the eighteenth century Isaac Newton, Leibnitz, and some other scientists defined the principles of mechanics. This advent in mathematics suddenly generated the Industrial Revolution of the nineteenth century where people became able to produce mechanisms, to introduce mechanical power that replaced human and animal power, to produce a revolution in the use of energy.

Since the Industrial Revolution, the rate of growth in human society increased exponentially at 2 percent - 3 percent or more.

However, the latest revolution in our society occurred in the second half of the twentieth century when artificial computation and electronics were introduced to assist the previously developed mechanisms and their functioning; meanwhile, the Internet was set up at the end of this period.

One of the most spectacular advances to occur was the introduction of smartphones almost fifteen years ago. This advance allows individuals to use their writing/reading skills in a much more extensive manner.

The last two hundred years produced the most radical changes in our society, butonly the changes in the last twenty years or so would be able to alter the architecture of the human brain further.

Since the nineteenth century basic education has undergone exponential growth as well. It accelerated the rate of slowing-down of perceptual processing, while it increased the revealing of details and options exponentially.

The most exponential growth manifests in the last twenty years when the amount of information produced every year is larger than the amount of information produced in all of human history. Every year we encounter a doubling of the information made available to human use.

THE SIDE EFFECTS

However, this exponential revolution in manufacturing manifested since the Industrial Revolution comes with plenty of side effects never experienced before. It produces an exponential growth in stress and changes human attention patterns. Stress, it seems, is the leading factor in all human diseases.

Gail Kinman, professor of occupational health psychology at the University of Bedfordshire (UK), found a significant increase in the perception of stress between 2006 and 2012, but there was also the sense that much more went unreported. She concluded that 2/3 of British employees agreed that people in their organization would be unlikely or very unlikely to reveal that they were experiencing stress-related illness.

In 2014 or 2015, a diagnosis of stress in the UK accounted for 35 percent of all work-related ill health, and 43 percent of all working days lost due to illness. The main work factors cited by respondents were workload pressure, including tight deadlines, too much responsibility and lack of managerial support. The evidence suggested that these problems are only getting worse.

There is little evidence that the rate of change in technology becomes the main stressing fact.

In a study performed recently, it was found that from 2000 to 2014, human attention dropped from twelve seconds to eight while goldfish attention is ranked at nine seconds. In other words, in less than fifteen years, human attention that processes the perceived information into our brains has lost one-third of its capacity due to the impact of the modern digital lifestyle. It was estimated that the current exponential evolution of Information Technology would contribute to a further drop of attention to around four seconds by 2025-2030.

The human capacity to process perceived information will become so tiny that the possibility to respond to harming and other interferences will become insufficient for unassisted survival. Most likely, it will affect our decision-making ability during normal life events.

A reduction in the acuity of human senses especially vision, hearing, and smell occurs. This reduction produces a partial impairment of these senses that inflicts an impairment in cognition. For example, the case of epidemic myopia in East Asia is well-known. In a study at Southampton University, Dr. Christopher Martyn said:

> What we found was very clear. The larger a person's head, the less likely their cognitive abilities are to decline in later years. In the opposite direction, persons with smaller heads have a fivefold increased risk of suffering cognitive decline and memory loss, when compared with those with the largest heads.

Another aspect refers to the human capacity to analyze and synthesize contents that rapidly increase. But in the opposite direction, it lowers our capacity to understand context, and this last fact may become, in some cases, a life-threatening concern.

However, the immense accumulation of data may become subject to sensible and dangerous thresholds.

DEHAENE THOUGHTS

In a recent interview with Gareth Cook, editor of *Mind Matters*, Stanislav Dehaene, chair of experimental cognitive psychology at the College de France said,

> What I am proposing is that the human brain is a much more constrained organ than we think and that it places strong limits on the range of possible cultural forms. Essentially, the brain did not evolve for culture, but culture evolved to be learnable by the brain. Through its cultural inventions, humanity constantly searched for specific niches in the brain, wherever there is a space of plasticity that can be exploited to "recycle" a brain area and put it to novel use. Reading, mathematics, tool use, religious systems - all might be viewed as instances of cortical recycling.
>
> A fascinating discovery, made by American researcher Mark Changizi, is that all of the world's writing systems use the same of basic shapes and that these shapes are already part of the visual system in all primates because they are also useful for coding natural visual scenes.
>
> The brain did not have enough time to evolve for reading - so writing systems evolved for the brain.
>
> In the case of reading, we now have plenty of evidence that the whole-language approach has nothing to do with how our visual system recognizes written words. Our brain never

> relies on the overall contours of words, rather it decomposes all of its letters and graphemes in parallel, subliminally and at high speed, thus giving us an illusion of whole-word reading. Experiments even suggest that the whole-language method may orient learning towards the wrong brain region, symmetrical to the word-forming area in the right hemisphere.
>
> The brain is a finite system, so although there are overwhelming benefits of education, there might also be some losses.

What this interview shows is that the same area of the brain is used for visual perception of scenes as well as for writing and reading. This fact allows one to presuppose a particular action of suppression that takes place in this area, which cancels the fast dynamics of the vision of scenes to the benefit of the much slower dynamics of writing/reading.

MUSIC EMBEDDED INTO LANGUAGE

The case of music therapy performed under the professional observation of teams of scientists from several universities showed significant effects of music on our brain functions such as improved memory and reading skills, increased blood flow to the brain, reduced stress and depression, and strengthened executive functions.

According to the Duke University team, the effects mentioned above might trigger the general stress reduction produced by music. Dale Purves, a professor at Duke University, discovered that the sonic range that matters and interests us the most is identical to the range of sounds we produce.

I will quote several ideas from an article published by Purves in June 2008 in *Duke Magazine*.

> The most salient periodic sounds humans hear on a consistent basis are the vocalizations we make to communicate with one another by means of speech. And speech, like music, is tonal. The tonal quality of speech is specifically associated with the production of vowels; this is so because our vocal cords only vibrate during vowel production.
>
> The Duke team hypothesized that speech played a major role in shaping the evolution and development of the human auditory system, including the tonal preferences found in music. The vowels and the frequencies they produce are those vocalized.
>
> Much of music's power lies in its ability to communicate without words, to speak directly to our emotions. Perhaps when we hear music on a major scale, we unconsciously associate it with happy speech and tend to have that emotional response, and vice-versa for music on a minor scale.

A UCLA study found that when one watches the performance of an athlete, the neurons which are associated with one's muscles will fire. Our muscles do not move, but the neurons act as if one is mimicking the observed. This type of sharing is a form of communication. A similar mirroring, sharing, and communication are at work when one listens to music.

THE CONTINUUM RESULTS FROM COMPRESSION BUT ALSO FROM DECOMPRESSION

Dr. D. M. Eagleman's essay, *Brain Time* (2009), states,

> The days of thinking of time as a river are over. Time perception, like vision, is a construction of the brain, and is shockingly easy to manipulate experimentally.
>
> After all, your hearing and your vision process information at different speeds. The brain constantly recalibrates its expectations about arrival time: all feedback should be assumed to be simultaneous, and any delays should be adjusted until simultaneity is received. In other words, your brain assumes that sound, sight, and touch are simultaneous. For the brain, this is critical to solving the problem of causality because causality requires a temporal 'order.'

It appears that the delay contains more details, what are they? We previously have seen that conscious states compress information. The decrease in speed seems to decompress some of that previously compressed information.

The slower speed also incorporates additional details. At slower speeds, more details of reality are revealed. These are perceived as details of the content.

Our brain selects and adjusts until the final product has a minimum of variables is displayed linearly and gives the illusion of order. This modality of work in our brains was triggered by the development of language that is essentially linear.

A group of researchers from Germany and Austria in the article, *Neuro-cognitive Mechanisms of Conscious and Unconscious Visual Perception from Phenomena to General Principles*, published December 1, 2011, in Advances in cognitive psychology explain:

> For unconscious conditions, processes triggered by unconscious visual stimuli fade quite fast, decaying within 200ms, because they are temporally less stable. In contrast, conscious visual representations are temporally very stable.

The authors concluded,

> We propose that conscious visual perception depends on a consolidation process within the visual system and a subsequent post- sensory gating process, which determines whether a stimulus is encoded in prefrontal working memory circuits. Only if both processes are successfully completed is the stimulus consciously perceived and reportable; otherwise, the stimulus remains unconscious.
>
> Attentional task sets were shown to influence unconscious semantic and visuomotor processes

> selectively, ensuring the adaptability of cognition even in the unconscious domain.

Consciousness is determined by encoding in memory. Whatever is not encoded in memory is unconscious. Deep-sleep and sleep-walking show that the procedural unconscious is active in such states, while half of the pyramidal cells remain active.

In the meantime, the brain spends 80 percent of its energy and metabolism in the resting state, while stimulus-induced activity requires only small incremental increases in the energy supply. Is this 80 percent of energy consumed by the pre-conscious and unconscious states?

Another quote from Dr. Eagleman:

> The more distant future of time research may change our views of other fields, such as physics. Our physical theories are mostly built on top of our filters for perceiving the world, and time may be the most stubborn filter of all to budge out of the brain.

Dr. Luder Deeke (Medical University of Vienna, Austria, June 2012) explains the role of consciousness and its origin.

> To have consciousness was a matter of survival. In evolution, brains became more complex, and to survive brain processing had to be fast: the fittest was a being that could act and react quickly. It only can be achieved using data reduction. Between the information flow in our senses and that in our consciousness, there is a selection; there is also unconsciously organized an enormous compression of information that

> takes place, data compression of several orders of magnitude (like 10,000 times or even more). It is the information compression that helps us to survive.

The question is: Would the decompression reflect what is called the difference in speed between visual and verbal that is assumed to be sixty thousand times or so?

The coding of information means compression, and decoding is the opposite process that decompresses information.

> There also takes place a pre-conscious filtering or selection/data reduction that occurs on a spinal level, but in the brain, too. Consciousness facilitates learning, selection, and freedom but also because it does not do everything alone it delegates most agendas to unconscious subroutines.

Our consciousness results from a selection of the brain to ensure survival. It selects and compresses data from our pre-conscious.

COULD I DEFINE THE SIZE OF CONSCIOUSNESS?

It is estimated that the conscious state represents 1–10 percent of brain activity; pre-consciousness seems to be 40–60 percent of the activity, and the unconscious is the rest or 30–50 percent.

It may be that the conscious state was under-developed in the era before thirty thousand years ago. Today estimates show that human perceptual speed decreased in this time interval of thirty thousand years by sixty thousand times. It means that, if the sum of the entire known reality thirty thousand years ago were equal to one unit, the modern current reality size would be sixty thousand times larger. It also implies that the size of the consciousness thirty thousand years ago was sixty thousand times smaller.

The magnitude of brain entropy is reduced by conscious processing by one million times.

The size and complexity of human consciousness have increased tremendously in the last thirty thousand years.

Why was it possible to add new information? This is because consciousness is a mechanism of survival. It has to select only a type of information that can be processed by the brain with only the computation available to it which is linear computation. However, the brain works in a nonlinear mode, using parallel processing, and this is reflected in unconscious processing. So, the unconscious is nonlinear. The information needs to present a minimum of variability to make sense to the conscious linear mechanism. The brain intervenes with filtering, compression, partial decompression, and other processes.

How does consciousness find new information? It reduces data to short time scales, and the slower speed alters a part of the pre conscious data that typically appears incoherent; by changing the speed, it becomes incorporated into the conscious mechanism as coherent—meaning less variability, less uncertainty. Gradually and directly proportional to the decrease in speed, more data becomes available for conscious states.

Marnix Naber, Olivia Carter, and Frans A. J. Verstraten in a paper titled *Suppression Wave Dynamics: Visual Field Anisotropy and Induced Strength*, published in *Vision Research* (by Elsevier, July 2009) explain:

> Evidence of visual field anisotropies in wave propagation speeds were also seen with suppression waves decelerating as they travel toward the fovea of the eye. The suppressed image will generally breakthrough, and its dominance will spread.
>
> After a hundred milliseconds, a mask pattern consisting of small moving dots is suddenly present to both eyes. Even though the masked pattern does not overlap the spatial location of the targets, they are perceived to disappear for periods of seconds. The latency of perceptual suppression increases with the target's distance.
>
> One means is unknown, by which sensory information is temporarily rendered inaccessible to consciousness, and it has neural correlates of unconsciousness."

> However, there are many other methods known to render stimuli consciously unavailable. This study will shed light on why and how stimuli are rendered unconscious. Here, there are several factors, which are constraining the transitions between perceptual awareness and suppression.

Another paper titled *Factors Influencing the Latency of Simple Reaction Time* (SRT) published in *Frontiers in Human Neuroscience* on May 26, 2015, the authors conclude:

> The slowed simple reaction time (SRT) in recent studies reflected a systematic reduction in processing speed, and hence fluid intelligence, in contemporary populations. But our study found that SRT's latencies have been overestimated over the past several decades.

MONKEY'S BRAIN

David A. Leopold, Yusuke Murayama, and Nikos K. Logothetis, in a paper published April 1, 2003, in Cerebral Cortex, showed that the signals in monkey visual cortex indicate fluctuations over time-scales of seconds and minutes. In general, they found fluctuations at many time scales, with a particularly large amplitude at very low frequencies (less than 0.1 Hz). These fluctuations exhibit high coherence. "These results demonstrate the existence of widespread coherent activity fluctuations in the brain of the awake monkey over very long time-scales." Response latency measurements, for example, suggest that visual stimuli are processed by a highly complex, parallel and recurrent cortical network (Bullier and Nowak, 1995; Schmolesky et al., 1998; Schroeder et al., 1998; Lamme and Roelfsema, 2000). Other studies have argued that the flow of information through this network is inherently multidirectional (Hupe et al., 1998; Lamme et al., 1998).

Anandamohan Ghosh, Y Rho. A. R. McIntosh, R. Kotter, and V. K. Jirsa published a paper in Computational Biology on October 10, 2008, from which I can summarize, "However, even at rest, in the absence of overt goal-directed behavior, collections of cortical regions consistently show temporary coherent activity. In humans, these resting state networks have been shown to greatly overlap with functional architectures that is present during consciously directed activity. In monkeys, similar coherent fluctuations are present during deep anesthesia, when there is no consciousness."

The monkeys seem to have a very penetrable boundary between their conscious states and their unconscious states. In other words, the monkeys previously accessed their unconscious states more frequently than humans did, but they cannot transform the unconscious data into coherent outcomes, as humans do.

They, like all other animals, jump from conscious certainty to unconscious uncertainty. We have domesticated this behavior in ourselves, and try to do the same with some animals and plants.

There are no studies on plants' unconscious behavior.

It seems that the monkey consciousness is much narrower than the human consciousness. The coherent states may rapidly dilute into unconscious and incoherent states.

"We specifically demonstrate that noise and time delays via propagation along connecting fibers are essential for the emergence of the coherent fluctuations of the default network."

"The combination of anatomical structure and time delays creates a space-time structure in which noise and intermittent neurologic oscillations enable the brain to explore various functional configurations representing its dynamic repertoire."

LANGUAGE HELPS TOOLMAKING

Thomas Morgan, a psychologist at the University of California in Berkeley, conducted an experiment in which he proved "only a group in which gestural or verbal teaching was allowed performed significantly above the reverse engineering baseline on several indicators of toolmaking skills."

The conclusion was that language might have helped early humans to learn to make advanced stone tools. He showed that "humans would learn more effectively when they could talk to each other."

Kolodny and Edelman show that "rudimentary language, which evolved in the context of toolmaking and teaching, was ultimately able to break away from its immediate contexts, eventually employing those original cognitive pathways for its own unique purposes."

In 2014, Chomsky and Robert Berwick said, "Language was acquired not in the context of a slow, gradual modification of preexisting systems under natural selection but a single, rapid, emergent event."

Mark Changizi (2011) said, "Language is a technology, not part of our nature at all." Language produced mental imagination, and this imagination invented tools.

THE ROLE OF REWARDS STIMULATED BY LANGUAGE LEARNING

There exists sufficient evidence that all biological systems have developed particular attractiveness toward various types of the high-rewards gains. In the natural world, there is embedded a natural competition for higher-rewards as a form of beneficial adaptation to environmental conditions and the condition change.

The evolution of human language proved to stimulate this type of natural competition because it introduced the learning of language from the earliest age. Hence, since their infancy that corresponds with their brain formation stage, the children were stimulated, benefiting from the role applied by the reward for their learning of words. It allowed the Paleolithic brain to enter a mental self-competition for high-rewards activity.

There was the memory that gradually developed better recording recognition for high versus low reward condition.

I have to mention that amygdala combines different sensory inputs. It combines internal and external stimuli, like the hippocampus, allowing many scientists to assume the existence of a connection between these with systems of the brain.

Emotions (like those of fear versus those of happiness) are negative and positive, while they seem to populate the extremities of this emotional spectrum. The middle is populated with neutral ones, which can fall into one basin of attraction or another. Both extreme types of emotions generate particular stress that activates the long-term memory recording mechanism. It appears as another mode of adaptation that is driven by stress.

However, the emotions are the extreme descriptions, which are made by a context-evaluating mechanism. In other words, the emotions are part of the environmental-stress that can be exterior or interior. As all contextual features, the emotions are nonlinear features.

The studies on the contemporary primitives indicate these people are very emotional. Our Paleolithic antecessors displayed the same emotional behavior. It was their response to the small and large environmental changes, some occurring to the outside world but others into their inner one.

Today, there is assumed that the prehistoric people used a distinct type of understanding than we do, where they combined all sensory activity to define the surrounding context. One might define this as a type of prehistoric intelligence.

We can estimate today that human consciousness evolved recently (last thirty thousand years) toward gradually but also fast switching the interest from context to content.

We can say today that the emotions were the basics of the prehistoric human thinking. It was that time before we started to delve into the content of things, breaking the nonlinear wholes into parts (where the sum of parts is not equal to the whole), and recombining such parts into new wholes, which become linear (where the sum of the parts is equal to the whole).

Nevertheless, there is known that the emotions have a higher connection to unconscious than any other thoughts.

We know that the unconscious shows a hierarchic display. An interesting theory states that chaos has a hierarchic display, where the separation is ruled by distinct degrees of randomness. This chaos has an edge toward our world of "order" that is a region of the instability of dynamic interplay between order and disorder. We intuitively know that this region is "creative" because here new patterns of "order" emerge from chaos, while these patterns are novel and different than the already known ones.

Returning to the hierarchy, it should be said that it defines a multi-scalar system, where on each scale are particular properties. It displays probabilities, as uncertainties, and which go from high to low to extremely rare, and to no-probabilities (unpredictability) that is pure randomness. A hierarchy of increasing degree of randomness is called ergodic.

But on the edge of chaos, the probabilities run from low to high. Probably, there is here a mechanism that increases and decreases their variability from high to low, and where the lower variables can enter our world of "order," where everything shows low variability. The changes in the hierarchical structure are perpetual.

As it appears, all hierarchies have a tree structure that also shows a "slaving principle" that focuses on constraint from a higher level to a lower level and where the lower level dynamics follows the higher level dynamics. The lower level variables can be eliminated, and the system can be mathematically written only by high-level variables.

We do not know if this mechanism acts directly on the behaviors of entropy, changing it from high to low or what else is happening up there.

The prehistoric world of emotions had a higher variability than our world of logic thinking. It was less stable, but in the meantime, it showed a high understanding of the contexts occurring in the environments. The emotions were a direct mental response to the changing contexts.

Since the evolution of language led thirty thousand years ago, to the creation of verbs and other grammatical forms (like nouns and adverbs with verb's functions), which all were able to code, discuss, and direct the action. They also reinforced natural sensibility toward the high-rewarding.

Emotional and motivational are both contextual where emotional characterized a more primitive phase, and motivational indicated a sudden jump in use beginning thirty thousand years ago. Both exacerbate a natural form of competition that gradually evolved into agricultural practice and the competition of the market place.

The fMRI activation provides evidence that the neural substrate of reward and motivation become active during word learning.

TWO HYPOTHESES

Scientific American on June 12, 2014, indicated that "MRI scans show that after an eight-week course of mindfulness practice, the amygdala appears to shrink. This primal region of the brain, associated with fear and emotions, is involved in the initiation of the body's response to stress. As the amygdala shrinks, the prefrontal cortex becomes thicker; the connection between the amygdala and the rest of the brain becomes weaker, while the connections to attention and concentration become stronger. Mindfulness recruits and increases higher-order."

A Dutch cognitive psychologist, Lorenza Colgato, in 2012 discovered that "open-monitoring meditation was far more effective in stimulating divergent thinking that is a key driver of creativity." On the other hand, the "focused-attention meditation was more strongly related to convergent thinking."

The above two examples may indicate the path human thinking took in the last thirty thousand years, evolving from emotional and motivational manifestations, which were contextual-visual-nonlinear toward logic-analytic-synthetic thinking that deals with the content and are verbal-linear.

However, the evolution was from a thin conscious layer that was having frequent unconscious interferences toward thickening of this layer by gradually adding more and more of that "orderly staff" that currently populates our modern world.

All these transformations have been intermediated by the revolution that occurred in language thirty thousand years ago.

In *Daily News* of June 26, 2016, an article was written by Claire Wilson about recent research at the University of Sussex in Brighton, UK, the author explains:

> Typically people who meditate feel like they decide to press a button about 200ms before their finger moves - but the electrodes in the non-meditating group reveal activity in the part of the brain that controls movement, and it occurs a further 350ms before they feel they make that decision. It suggests that, in fact, it is the unconscious brain that 'decides' when to press the button.

Here, it appears a connection between emotional-brain and the meditation-brain, and it manifest as an unconscious neural substrate. It may show that the practice of meditation—voluntary or involuntary, played an important role in modern brain evolution. There is to consider that the unintentional and intentional use of plants with hallucinogenic effects was sufficiently extensive in the prehistoric times to produce permanent changes in our brain architecture. The paramount beneficiary was represented by significant increases in human attention, incipient production of creative behaviors, and the newly manifested curiosity toward "content."

Especially the attention had led to plenty of "discoveries," which before were hiding in plain sight.

MEDITATORS DEAL WITH THE UNCONSCIOUS

In a new study, "The meditators had a longer gap in time between when they felt like they decided to move their finger and when it physically moved - 149ms compared with 68ms for the other people."

"This suggests they (meditators) were recognizing their unconscious brain activity earlier than most people." This study suggests that the time interval necessary to access the preconscious/unconscious manifests like a gap of 150ms (350ms minus 200ms). The conscious reaction appears with a latency of 150ms versus the original reaction of the preconscious/unconscious. For meditators, this time interval is reduced to 81ms.

Now, this same paper describes another experiment, where the non-meditators would be hypnotized several times. After the hypnosis sessions ended, the subjects were tested again. "Those who would be easily hypnotized felt like they decided to move the finger 124ms later than did those of low hypnotizability." In fact, the easily hypnotizable group had the sensation of deciding to move 23ms after their finger had actually moved. So, as it seems, "the people who are highly hypnotizable may have less conscious access to their unconscious intentions" because they need an additional 101ms to make this access possible.

Hypnosis is a process that opens up the unconscious information, but mostly in the form of some unconscious memory. Here, the discussed experiments suggest that opening the door to unconscious data during hypnosis requires a forcing factor, for example, the person who hypnotizes the subject. It is like manually opening a door for individuals who cannot allow this process to occur in an automatic mode. In contrast, other individuals, like those less able to be hypnotized, have an automatically opening the door toward unconscious data, while they refuse the manual type of opening. The manual mechanism becomes blocked by the automated one.

The meditators are training themselves to open the door to the unconscious automatically. This training leads to significant increases in attention, and there is the aspect of that "unconscious attention" which has been emphasized.

Stephen Fleming with WTCN of London suggested that people who meditate are less easy to hypnotize, and people who can be hypnotized are less mindful or less aware of their internal bodily processes.

People who meditate are more aware of their brains' unconscious activity.

THE SIZE OF CONSCIOUSNESS

The scope of the above information refers to the manifestation of latency between the speeds of conscious activity and the preconscious/unconscious processing speeds.

This delay varies from 150ms to 101ms and 81ms, depending on the characteristics and behaviors shown by each subject.

This latency is similar in size to the saccadic motion latency manifested during spoken language production. Both are about 150ms. It may be an indication of the size of our conscious activity: it might be placed at a speed-distance of 150ms from the preconscious/ unconscious activity. Spoken language introduced the same size delay: does it mean that the same latency—150ms, is the sole factor that produced language and also human consciousness?

This number—defining the latency as the thickness or depth of consciousness, seems to be 150ms for the average human.

In a study mentioned at the beginning of this paper, it was found that monkeys show a latency that is 20–30ms shorter than in humans. The said value of 20–30ms must be the distance traveled by evolving human consciousness from before 100,000 years ago to the current epoch. So, the monkeys' conscious activity is 20–30ms thinner than that of humans.

In the case of mental illnesses, the delay/latency may increase to 300–500ms. Could it be that the boundary between the conscious and preconscious disappears? Is 500ms the thickness of consciousness plus the preconscious?

Now, what exactly is in the preconscious/unconscious? They are expressed by different rhythms in the brain like Alpha, Theta, and Delta. Conversely, consciousness seems to be defined by other rhythms, like Beta and Gamma.

The preconscious/unconscious seem to represent a kind of massive mental library, where there are stored all memories, habits, behaviors, etc. It is everything one has learned and experienced in a lifetime.

The conscious mind has a developed data filter. During our daily experience, we accumulate some two million bits of information per second. It is the role of the filter that selects a manageable amount of information that becomes our consciousness.

What exactly would such a filtering mechanism do? Besides its capacity to compress data, it separates the information placed on some time-scales from information placed on other time-scales. The size of the time-scale givesinformation an orderly (low variable) or a random (high variable) format.

Our world, as we see it, seems to be a world of order, and the conscious activity and its filtering select this order.

LANGUAGE AS TECHNOLOGY

In a paper in September 15, 2016, titled *Language Could Be Humankind's Most Impressive Technological Invention*, Salikoko S. Mufwene, distinguished service professor of linguistics at College University of Chicago, explains:

> Languages are comparable to other technologies humans have invented to survive more easily... An important difference is that languages are technologies, which use our own anatomies. Spoken language was developed by co-opting

> the anatomical structures for both breathing and chewing to articulate various speech sounds.

Language generated culture but also is an embedded aspect of culture "assuming that culture consists of, roughly, the particular ways members of a population do things and behave, including their beliefs and how they interact with each other and their natural ecology."

Co-opting human anatomy, we created language as the most important technology we ever invented. All other technologies we have invented are all subservicing the primordial one that is language.

As time passed, language evolved into a better fidelity communication system. In a way, all technologies we ever made, but mostly in the last seventy thousand years or those being more relevant, which have occurred in the last forty to thirty thousand years have to serve language as a paramount means of communication.

We have to assume that sign language, which evolved in parallel with tool making, was also a technology derived from our anatomy. It gradually, but very slowly advanced into vocalization.

MATHEMATICS IS LIKE A SECOND LANGUAGE

Not many people see that mathematics is an intrinsic part of the language. *Literacy Strategies for Improving Mathematics Instruction*, published by Joan M. Kenney, Euthecia Hancewicz, Loretta Heuer, Diana Metsisto, and Cynthia L. Tuttle.

> A model proposed... at the Harvard Graduate School of Education (Schwartz & Kenney, 1995) suggests that we think about mathematical nouns, or objects, as being numbers, measurements,

> shapes, spaces, functions, patterns, data, and arrangements - items that comfortably map onto commonly accepted mathematics content strands. Mathematical verbs may be regarded as the four predominant actions that we ascribe to problem solving and reasoning: modeling and formulating, transforming and manipulating, interfering, communicating. Taken as a whole these four actions represent the process that we go through to solve a problem.

All in all, mathematics is a languagewhile truly this is a foreign language for most students. Mathematics is not our first language.

However, in self-consciousness, the "self" is described by the language; it is the hero built by the language.

THINGS BEYOND CONSCIOUS AWARENESS

As Seymour Epstein discovered in 1973:

> We have the ability to understand and even quickly learn things beyond conscious awareness." (cited by Wilson 2002, 2012, 2014, Kahneman 2012). "This experiential system processes percepts, not concepts, and so its operations remain unconscious and always hidden from our conceptual understanding.

In my opinion, these things above are the details or options triggered by the latency of the saccadic movement. They are things from the unconscious mind, which suddenly or gradually become exposed. Of course, they are only information.

The crowding of novel options seems to be the main stressor for brain functioning.

This increased latency in the saccadic movement may produce a lot of unwanted effects. It appears to be so because all mental illnesses have in common unusual increases in latency that are supposed to generate more options than language alone can generate. This unusual crowding of optionons would generate great confusion and may inflict an alteration in the functioning of the brains of mentally ill subjects.

Most specialists agree that language is essential for higher levels of consciousness such as self-consciousness and meta-consciousness (Carruthers, 1998; Neuman and Nave, 2010). It also becomes obvious that lower-level consciousness such as general consciousness and self-awareness do not rely on language because they exist in verbally incompetent beings and preverbal infants.

To access the content of consciousness and cross-modal attention, one requires a language. Social behaviors increase in complexity because of language. Rational capacities increase with language. Language allows one to make his/her thoughts explicit, to ask for reasons and justifications as a base for their knowledge.

Language is a necessary condition for achieving an advanced consciousness. It also seems that our linguistic capacities help create structured thinking. The language also helps to report various conscious states.

SIGHT UNSEEN

As we have discussed before in this paper, there are biases of unconscious selective attention. Some events or objects become visible or invisible based on what they mean to you.

In a paper published by the American Psychological Association (April 2001, Vol 32, No.4) the authors explain:

> Intuitively, people believe that as long as our eyes are open, we are seeing." "But increasingly, studies of visual perception have demonstrated how startling little people see when we're not paying attention, a phenomenon known as "inattentional blindness." "We can fail to perceive very major things going on right in front of our eyes," remarks cognitive psychologist Brian Scholl, Ph.D. of Yale University.
>
> Beginning in the '70s, the researchers also defined a phenomenon named "change blindness" that refers to the fact that people fail to recognize and detect "changes in their visual field.

From these researches arose several major questions such as (1.)"How much input can the mind encode—consciously and unconsciously?" (2) "What brings some visual objects to conscious awareness, while others remain unnoticed?" and (3) "What is the fate of information that is perceived only unconsciously?"

In 1998, psychologist, Arien Mack, and Irving Rock of the University of Californiain Berkeley concluded that the brain undertakes considerable perceptual processing outside of conscious awareness before the attention is engaged and that objects or events that are personally meaningful are most likely to capture people attention. Inattentional blindness may depend on the similarity between unexpected and attended objects: participants were more likely to notice a gorilla, whose fur was black when they were attending to the basketball team in black shirts than when they were attending to the team in white shirtsAt any given time, people may be attentionally tuned to certain perceptual dimensions, such as brightness or shape. Most people are really focused on one goal at the time.

> The results suggest that attention is critical not for engaging perceptual processing but rather for encoding the products of those processes into short term memory so that they can be reported.
>
> One conclusion of the inattentional blindness work has been that we see far less of our world than we think we do. We might encode much of our visual world without awareness.
>
> We can't tell whether everybody experiences inattentional blindness about 30 percent of the time or 30 percent of people are characteristically inattentional blind.
>
> Early results suggest that inattentional blindness is actually part of a more general sensory phenomenon, where it extends to hearing and touch.

However, our attention work is to define patterns in the surrounding environments. The fundamental function of the brains of all animals is to encode and integrate information acquired from the environment through sensory inputs, and then generate adaptive behavioral responses.

As appears, the main work of all brains is to process such patterns. We, as humans, are fond of higher or superior processing of these patterns.

But what are these patterns?

At the quantum level, we have a discontinuous motion of particles that takes place in a continuous macro space-time. It means that one has to anticipate the collapse of the microscopic wave function that brings at the macroscopic scale the continuous motion of objects.

The research over chaotic phenomena unveiled the discontinuity of that space where the chaotic phenomena take place.

It still is subject of debate the possibility that the discontinuity separates one pattern of another. It means that a pattern is coherent only inside a space with continuity; but here, the coherent behavior is emphasized only when it is less fluctuating and; hence, it has a reasonable degree of uncertainty. The size of that space does not matter.

Those eleven million bits of information which our oculomotor system perceive every instant are subject to a very drastic selection that allows only 40 bits per second to become effectively used by the processing of the neuronal network. It is assumable that most of those 40 bits describe information about some patterns.

As talked before, these forty patterns have the lowest variability. By contrast, these patterns show slow changes, which allows one to assume some predictable features about their future evolution. They also show a finite evolution, display symmetry, repeatability and proportionality. This means one can find every single pattern repeating or replicating itself time and again within the same range of linearity.

From these 40 bits/second some are subject to our attention (probably some 60–70 percent), but a certain portion (30–40 percent) will escape this processing. We deal with no more than 25–30 bits in a single instant of time.

In my opinion, the patterns that escape our attention can become observable when one changes the focus of his/her attention. They become what we call our great discoveries, but here the change in the attention focus is a mental construct that introduces a new purpose. It is like we will see what we like to see.

I would say that these 40 bits are the occurrence probabilities which are generated at an intersection of probabilities. Such intersections turn low probabilities into higher and occurrence probabilities. Such higher probabilities increase the precision of prediction, producing the correction of the prediction errors.

It is important to understand that patterns and processes observed at finer scales are those operating over broader spatiotemporal scales and similarly, how large-scale processes correspond to small-scale phenomena.

The discontinuity approach is derived from hierarchy theory: each level in this hierarchy is like nested and represents variables processes which are sufficiently diverse to generate discontinuities between regions with different behaviors. Here, each such region of space is dominated by a particular behavior (being it speed, size, morphology, attributes, etc.) and every such region defines a "pattern."

SUMMING UP AND MULTIPLYING PROBABILITIES VERSUS MIND WORKINGS

The mathematics of probabilities indicates the following possibilities.

The addition of probabilities only works when the events to consider (here the events are being the probabilities) are mutually disjoint—meaning that no intersection occurs. In the case mentioned above, where the events do not have any common outcome, the sum of the probabilities means their "union."

The product rule (or their multiplication) also refers to independent events and the intersection equals the product of probabilities.

The computation of probabilities takes place in all animal brains. However, our understanding of the world around us is imperfect, and when dealing with "chance," our brain tends to come up with ways to cope with the unpredictable nature of the world.

Michael Shermer on September 1, 2008, explained in his article, *Why Our Brains Do Not Intuitively Gasp Probabilities,* published in Scientific American:

> Let us assign a number of one bit per second to the data that flows into our senses as we go about our day and assume that we are awake for 12

> hours a day. We get 43,200 bits of data a day, or 1,296 million a month. Even assuming that 99.999 percent of these bits are meaningless (and so we filter them out or forget them entirely), that still leaves 1.3 "miracles" a month, or 15.5 "miracles" a year (miracles are here outlier anomalies).
>
> Thanks to our confirmation bias, in which we look for and find confirmatory evidence for what we already believe and ignore and discount contradictory evidence, we will remember only those few astonishing coincidences and forget the vast sea of meaningless data.
>
> These and other examples show the power of probabilistic thinking to override our intuitive sense of numbers. Folk numeracy is our natural tendency to misperceive and miscalculate probabilities, to think anecdotally instead of statistically, and to focus on and remember short-term trends and small number runs.

Olivia Goldhill, on March 4, 2017, expressed in the article, *Humans Are Born Irrational, And That Has Made Us Better Decision-Makers,* published in Quartz:

> In the 1970s, two psychologists proved, once for all, that humans are not rational creatures. Daniel Kahneman and Amos Tversky discovered "cognitive biases," showing that humans systematically made choices that defy clear logic."
>
> Despite the growing reliance on "big data" to game out every decision, it's clear to anyone

> with a glimmer of self-awareness that humans are incapable of constantly rational thought. We simply do not have the time or capacity to calculate the statistical probabilities and potential risks that come with every choice.
>
> We live in a world of deep uncertainty, in which neat logic simply isn't a good guide.

Gerd Gigerenzer, director of the Max Plank Institute for Human Development in Germany, who studied decision-making in real-world settings says:

> There is a major flaw in any system that attempts to be overly rational in our highly uncertain world.
>
> In a world where you can calculate the risk, the rational way is to rely on statistics and probability theory. But in a world of uncertainty, not everything is known-the future may be different from the past-then statistics by itself cannot provide you with the best answer anymore.

Henry Brighton, a cognitive science and artificial intelligence professor at Tilburg University in the Netherlands, adds:

> The number of objective facts deserving of that term is extremely low and almost negligible in everyday life. The whole idea of using logic to make decisions in the world is to me a fairly peculiar one, given that we live in a world of high uncertainty which is precisely the conditions in which logic is not the appropriate framework for thinking about decision-making.

Gigerenzer also said:

> In a world of uncertainty, heuristics (as instincts) tend to be more robust. We need hot-headed (overly emotional), emotional decisions.

"Chance," Kahneman writes, "is commonly viewed as a self-correcting process in which a deviation in one direction induces a deviation in the opposite direction to restore the equilibrium. In fact, deviations are not corrected as a chance process unfolds; they are merely diluted."

HOW WE CAME TO INVESTIGATE PROBABILITIES

The prehistoric humans were heuristic: their instincts and emotions guide them. The emotions are the result of changes in the context.

Emotions are carried out by the limbic system of the brain that is located in the temporal lobes. For example, the removal of amygdala severely alters, diminishing the emotional response.

The hippocampus is another part of the limbic system that sends information to the amygdala. In the meantime, the hippocampus is one of the memory processing centers of the brain: the emotional memories are sent to amygdala for processing.

The prefrontal cortex is the decision center, including those based on emotional reaction.

Hypothalamus is another part of the limbic system that sends information to the amygdala. It regulates the emotions.

Cingulate gyrus is a pathway involved in emotions that connect the thalamus with the hippocampus. It mostly deals with *paying attention to various events*, which events might incite emotions.

According to with several studies, psilocybin mushrooms work by decreasing the activity in frontal lobes, while increasing the activity in other areas, like the emotion-processing areas, which because of recent evolutions have turned significantly less active. The psilocybin acts over the blood flow in the brain while the blood is the main supplier of oxygen for neural activity.

Professor David Nutt, from the department of medicine at Imperial College London, who was the senior author in both mentioned studies explains:

> Psychedelics are thought of as a mind-expanding drugs so, it has been assumed that they work by increasing brain activity; but surprisingly, we found that psilocybin actually caused activity to decrease in areas that have the densest connection with other areas. These hubs constrain our experience of the world and keep it orderly. We now know that deactivating these regions leads to a state in which the world is experienced as strange.

This world of strange described by Professor Nutt is a world of higher variation that is opposed to the orderly behaviors we experience.

However, the experiments and the analysis of the psychedelics showed that the recorded effects refer to decreases and increases of local entropy of various brain sections; in other words, some areas experience more activity than others and this type of switching is opposed to the currently known type of activity.

The psychedelics, whatever their nature is, bring our brain to an intermediate phase between its prehistoric but emotional version and current modern but logic version.

This phase was developed at some point in our last thirty thousand years of evolution and it was responsible for calming our emotions (like a type of self-domestication), while on the other hand, it *stimulated and enhanced the attention driven processes*.

Surprisingly, the meditation corresponds to these same areas of the brain which are affected during the experiments with psychedelics.

The recorded short-term and long-term results are somehow different between meditation practices and psychedelics experience, but overall, they show very many similarities.

In my opinion, the phase mentioned above of the transitional brain contributed significantly to slowdown the emotional-contextual states, while gradually inserting new functions like those *developing the attention skills*.

When we mention the domestication of animals and plants, and later the agricultural practices, we already talk about the *first grand discoveries of the modern brain, which resulted from increased attention*.

The *attention* made people observe the details of the environment, which before this time, they were things which have always been ignored. This change in behavior, in fact, moved from a human preoccupation with the context (of the environment) toward a type of curiosity for the details of the content. It was the first step in delving into logic approaches: it was about analyzing the content of the reality, about breaking it into pieces and reconstructing it in a different manner (synthesis).

The verbs introduced the concept of and the motivation for action. Gradually, the contextual world of emotions and motivation was altered by the attention. It was the timeless world ruled mostly by the unconscious that gradually introduced the concept of the future, made the time flow like a river, and began *producing "discoveries" based on the new technology of "attention for details."*

It was a prehistoric world of "uncertainty" that gradually became more "orderly" in the newly evolving mind. The "prediction" was needed, and it was invented because the new world was based on "a certainty built in mind."

The new world became the world of "probabilities." It became a world rising higher and higher in mind while dissociating us faster and faster from the surrounding natural world.

SLOWER SPEED OF PERCEPTUAL PROCESSING

One of the essential parts of the new mechanism created in modern brains is the slower speed of perceptual processing that is generated by the latencies and delays in the oculomotor system.

It affects brain functioning in many ways, but all are interrelated: It generates a fine-tuning of the environment, bringing more details from the near-field and details are appearing as a matter of content. It is a fine-tuning of decoding of information.

It brings from the unconscious more information that usually cannot be perceived. It makes humans programmable. It increases the efficiency of neurons at slow speed interactions, making slow learning more efficient than fast learning.

In general, the new slow speed of processing had to favor the development of a verbal mode that is much slower than visual modes. The slow speed of processing favored analytics and mathematics. Analysis of content becomes predominant in human judgment and activity and leads us to disassemble wholes into parts, while the parts are rearranged in multiple ways.

The slower speed of processing favored writing and reading which in turn favors an increase in latency. It lowers the entropy of the brain, bringing us an apparent world of order. It may be the case that because of the change in human diet, our brain was forced to adapt to new ways of processing, making the neurons more efficient at lower metabolisms.

It makes humans good learners but turns them into slow-speed-learners, as an adaptation to content-learning. It makes language the best technology we could have. In the end, all these things bring an immensity of details to our perceived reality, and this increased exponentially starting two hundred years ago, while the rate of increase became much more accelerated in the past few decades.

New characteristics favored by the new brain mechanism separated humans from the rest of the animals and made them unique, eventually making us what we are today.

IMPLICATIONS ON INTELLIGENCE

A psychology handbook states that

> Slow or poor processing speed is not related to intelligence, meaning that one does not necessarily predict the other. Slow processing speed means that some determined tasks will be more difficult than others. It may also interfere with executive functions, as a person with slow processing speed will have a harder time planning, setting goals, making decisions, starting tasks, paying attention, etc.

Beno Csapo and Joachim Funke in *Nature of Problem Solving* published in 2017, conducted extensive research on high school subjects from Germany (2003–2012) and concluded:

> Analytical problem solving (all the information needed is explicitly stated or can be inferred) competence is higher than intelligence. By

> contrast, dynamical problem solving (most of the information needed has to be generated or learned by doing) competence is lower than intelligence.

Csapo and Funke explain that analytical problem solving refers to a defined content, and here our analytical power is higher than our intelligence. But in the case of dynamical problem solving, we deal mostly with context and our competence is lower than our intelligence.

It shows that today high school students have a mind-set for analytics and contents, but when the case of context arises, their competence is much lower than their intelligence. Here competence refers to ability.

Their intelligence serves well only in matters of content.

Harvard Business Review published a study, *Teams Solve Problems Faster When They're More Cognitively Diverse* (March 30, 2017). The study concluded:

> Tackling new challenges requires a balance between applying what we know and discovering what we do not know that might be useful. A high degree of cognitive diversity could generate accelerated learning performance in the face of new, uncertain, and complex situations, as in the case of the executive problem.

This paper emphasizes the role of cognitive diversity in problem-solving; again, the role of intelligence is lower than the role played by the diversity of knowledge and practices that define the context.

Again, diversity is the context; it deals with parallel and nonlinear approaches, and intelligence capable of dealing with contexts is significantly more valuable than one dealing well with content alone.

In the paper, "Working Memory and Fluid Reasoning: Same or Different?" published on January 22, 2014, the author, Berry Kaufman, found, "In the case of the "highly speeded group" (20 minutes), working memory explained all of the variances in fluid reasoning, whereas in the "unspeeded group" (60 minutes), working memory accounted for only 38% of the variance in fluid reasoning."

When people speed up to solve a problem, their fluid reasoning will depend entirely on their working memory. When people have more time to decide because they process the information slowly, they will use fluid reasoning 62 percent of the time, and working memory only 38 percent of the time.

This last study confirms the results obtained by Dr. Sebastian Goldt (University of Stuttgart) that "the slower the neuron learns, it increases neuron efficiency."

In the last two decades many papers and analysis of the current level of human intelligence seem to describe a downturn according to some, and an upturn according to others.

We have experienced a process of slowing down of our perceptual speed since thirty thousand years ago. The main culprit and benefactor, in this case, is the development of human language.

The development of language generated new mechanisms in the brain, which are now part of the technology of language itself.

The situation is complex, and I could not say that according to this decrease in perceptual speed people become more stupid or less intelligent. I would say that, because of language, we have a different brain that functions differently.

In this new way, some of our neuronal circuits turned to work at lower speeds and there are multiple consequences:

some are pleasant but some are highly unpleasant or even detrimental or dangerous, like stress and illnesses.

The immensity of details that have been generated is beyond human computational capability. Here, we had very much information but we did not know where it led to.

We found a way to create an artificial mechanism to compute information by compressing it linearly. This artificial compression gives us a clue about the context, but it remains a linear view of the context.

BRAIN RHYTHMS

G. Buzsaki said:

> Despite the several-thousand-fold increase of brain volume during mammalian evolution, the hierarchy of brain oscillations remains remarkably preserved ... at the approximately same speed, irrespective of brain size.

It is important to indicate that preservation of cortical rhythms across species does not reflect the brain's inability to change timing mechanisms. So, the constancy of brain oscillations across species reflects the importance of timing. The brain's architecture is subordinated to a temporal organization. The preservation of frequency in the various mammalian species is an important argument in favor of time as the organizing principle of these brains' dynamics.

Energy is proportional to frequency, and it is transferred through vibrations of electrical and magnetic fields. Even chemical compounds show vibrations. The oscillations are developed between kinetic and potential energy. The greater the energy, the larger the frequency; but in this case, more amplitude equals more energy.

Electrical signals are converted by simple mechanism into sound waves. Human language evolved as a conversion of electrical signals of waves and rhythms through an anatomical device into sound waves in the brain. In the anatomical mechanism, sound propagates in solids and liquids as a longitudinal wave or compression wave.

Returning to the brain rhythms, slow and fast waves work together through their coupling. By coupling, many brain regions can be connected at the same time at different frequencies, but also one rhythm can nest within another. Networks use specific frequencies to assure information communication between various neuronal groups.

Information is encoded as a signal into the frequency modulation of the anatomical mechanism. The result is the sound of voice activation that is decoded by the listener in a reverse process. The voice activation compresses the information from the brain, and the listener decompresses it into the information that is processed further by his/her brain.

This sound-based mechanism alters somewhat a similar process that is induced by the visual encoding-decoding mechanism. Both have a wave and electrical activity of the brain in common, but a different speed of processing visual vs. verbal. Language evolved from the simple encoding of a few sounds into a much more complex system of encoding-decoding of narratives, when it began to produce not only some simply vocalized nouns but also introduced the means of action through the coding and decoding of verbs and adverbs. This new ability was associated with and reflected in sudden progress in tool making. It became possible to communicate from one individual to another some full procedures of production. For the first time, the means of action were introduced to the conversation beyond the meaning of gestures.

When one analyzes the brain's rhythms and other elements of neurophysiology, one finds that our brain does a colossal job of natural computation. By different means, the brain analyzes and communicates information from the environment and our body, coordinating our actions. This activity of the brain is sufficient to understand that galactic rays that come to us as cosmogenic radiation are in fact wave phenomena which transport energy and information.

By the phenomenon of resonance, this information can be passed to our neuronal circuits, eventually altering their functioning.

When a particular stressor or a group of stressors manifests in the environment, the organism tends to respond by adapting without genetic mutations. However, the persistence of the stressors may favor a change in information processing in the brain, and this is the mutation.

We are confident that all animals are subject to environmental stressors, and all of them are also subject to the same galactic radiation, but there is always a manifestation of the particularities of each species, and stressing details. It is a local factor that makes a geographic distinction between various localities.

These details and locations can make some factors seem to be more relevant for one species compared to another. And this can favor a mutation in one species but none in all other species.

This same thought would also apply to the case of humans and Neanderthals or Denisovans. Was the evolution of human language the paramount element that abruptly separated humans from Neanderthals? Or, is the language the element that turned Neanderthals into humans until a full but relatively fast absorption occurred?

A LANGUAGE UNIVERSAL

An article of Kazuo Ueda and Yoshitaka titled *An Acoustic Key to Eight lLanguages/Dialects: Factor Analyses of Critical-band-filtered Speech* published in Scientific Reportsin 2017, found the presence of a common element that suggests a language universal.

The analyses robustly led to three factors common to all languages investigated - the low & mid-high factor related to the two separate frequency ranges of 50-540 and 1,700-3,300 Hz, the mid-low factor in the range of 540-1,700 Hz, and the high factor the range above 3,300 Hz - in these different languages/ dialects, suggesting a language universal.

This research may indicate that human spoken languages have a common frequency generator mechanism in the brain, which produced the original encoding of vocal sounds in the language that seems to be the same in all languages. It may also show what would be the fundamental communicating message of the brain that was encoded in all languages.

CONCEPT OF "GOAL"

Every living being in existence is goal oriented but such goals are very specific, concerning food, procreation, self-preservation, and a few other less important scopes. Every creature has one goal or several at the time, but we do not know about them because we do not understand how these creatures communicate, or for what purpose these goals are selected.

Literature has a large selection of studies on this subject that range from the analysis of goals in non-humans—monkeys especially, to human infants; children and; adults.

In an article titled *The Unconscious Mind* (January 2008), John A. Bargh and Ezequiel Morsella of Yale University show that their research has demonstrated the existence of several independent unconscious behavioral guidance systems: perceptual, evaluative, and motivational. They found that all actions of the unconscious mind precede the arrival of the conscious mind, and where that action precedes reflection. There is known that *conscious is action and unconscious is reaction or reflection*.

The authors say:

> our genes have provided us not with fixed responses to specific events, but with general tendencies that are adaptive across local variations. Genes primarily drive our behavior

> through motivations. Evolution works through motives and strategies.
>
> Many recent studies have now shown that *unconscious "goal pursuit" produces the same outcomes that conscious goal pursuit does.* The "goal concept," once activated without the participant's awareness, operates over extended periods to guide thought or behavior toward "goal."
>
> The open-ended nature of such unconscious "goal" pursuit is revealed by the fact that the "goal" operates on whatever "goal-relevant" information happens to occur next in the experimental situation. (subliminal, of course), which could not be known to the person beforehand-just as our genes programmed us to be capable of adapting to and thriving in local conditions far into a future that could not be anticipated in any detail. That the unconscious operating "goal" can adapt to whatever happens next and use the information to advance the pursuit of the goal.

However, my concept of "goal pursuit" starts from the premise that human language generates a virtual-space latency in the oculomotor system that is in large part language-oriented; this means that the neural systems must fill up this latency with some information that will assure the continuity and flow of action during the conversation.

For visual related purposes, the oculomotor system latency must be filled with predictions about motion, having the same scope to provide a continuum of motion and time flow. During the human conversation, the same necessity to fill a latency arises. But here a fundamental question appears: what is the nature of this information filling? Most literature suggests that most of the information is goal-related.

Human language appears to the mechanism of the oculomotor system like a distraction that generates a latency that is the same size as, or even larger than, the size of latency generated by motion. The neuronal mechanism that deals with this virtual-space, language-related latency provides the anatomical distinction between humans and all non-speaking animals.

However, the nature of such goal-related information is fundamentally different from the rest of the animals because the unconscious provides it to a conscious state that is managed and structured by the language itself. Because of this, the nature of this goal differs.

David Premark, in the article, *Human and Animal Cognition: Continuity and Discontinuity,* published in PLOS in August 2007, tried to find a novel explanation for the differences manifested between humans and other animals. He found several fields where he said most differences exist. Among them were goal-orientation, language, and education. I believe that only the goal-oriented approach deserves a special look because the rest is a result of the evolution of language.

Human language crucially changed from static to dynamic when verbs and adverbs were introduced next to nouns and adjectives. It occurred some forty-three to thirty thousand years ago. This introduction accomplished several important things: it introduced the continuity in vocal expression, which strengthened the flow-of-time concept in the human mind, and it introduced the sense of "virtual action" to the language, verbs allow programming of action; but most of all, they introduce programming as a means of control.

As many scientists see, language reflects very little means of communicating while the most important task of language is the capacity to control behavior. When we talk about goal-oriented human skills, we have to introduce these new capacities of virtual dynamics, which were produced by human language around thirty thousand years ago and which turned language into technology for programming and controlling.

Human goal-orientation develops the programming of goals and these goals become the tools used to control each other, the ecosystem, and the environment.

Thus, human goals become distinct from the goals of any other animal because they have the character of control-oriented-goals. They are a hallmark of prospective control.

Goal-oriented activity has a neural substrate of processing, and it becomes synonymous with the foundation used in decision making. Decision making is also goal-oriented, while together they accomplish an act of planning and controlling.

The language was static until the introduction of verbs and adverbs. This introduction generated a virtual space in mind, the space of language. This space became filled with goal-oriented information that was experimentally proven to be programming and control-oriented.

However, the mechanism presented above is also inflicted in producing "imaginary thinking." Here, the verbs became the vehicle that propelled the imagination, allowing the creation of dynamic outcomes. The noun is the part, which is recombined. But the verbs activated them.

SMALL, REPEATED, RECENT GENETIC CHANGES HAVE MANIPULATED THE WORKING ENTROPY OF THE HUMAN BRAIN

I will start by mentioning Steven Mithen's book, *The Prehistory of the Mind,* published by Thames and Hudson Ltd. in 1996 in the UK. After huge documentation, the author came to several very interesting and novel conclusions.

> To domesticate animals and plants, it was necessary for prehistoric minds to be able to think of them as beings with whom "social" relationships could be established. We can see evidence for the emergence of social relationships between people and wild animals and plants among the prehistoric hunter- gatherers of Europe. Psychologist Nicholas Humphrey drew attention to the relationships people have with plants because they bear close structural similarities to those with other people. They treat animals and plants as social beings, so too, they treat them as artifacts to be manipulated. It was a form of environmental management/ manipulation. People wanted to control the

> environment. People imagined this control, and focused on materializing this imagination.

The above quotes from Dr. Mithen's book describe an evolving social organization and the apparition of new relationships among prehistoric humans living in the Upper Paleolithic.

It is the epoch named by many researchers as the "Paleolithic Explosion" when the language and art got a sudden fast development of a very distinct quality compared with all other past developments.

The epoch started maybe sixty thousand years ago, but the progress was very small in the beginning. About forty thousand years ago, each step of progress became significant.

Here, the genetic change is evident. We have to assume a lot of small incremental changes, which are known to be much more successful than large changes. What exactly did the genetic change? There is enough evidence that the *changes manipulated the working entropy of the human brain*. In general, it occurred a lowering of the entropy that started to generate the "illusion" of "order." This "illusion" came in the form of a lot of "imagery."

This same "illusion" helped to develop the language much more rapid than in the previous ages. The role of language was to "establish control" because the world of "order" was based on "control." The language enlarged the consciousness.

Now several changes started to manifest in human thinking. The prehistoric thinking was "emotional" and is based on the experience achieved with the exterior world, so it was contextual and it was using all senses combined.

One of the basic elements of the newly created world of "order" was the language complexification where the linear rules of grammar begin to be implemented. The grammar transformed the language, and the language transformed the thinking. Languages are all linear, except one in Amazonia.

It could be said that in a relatively short interval of time the language begins to restructure the thinking, transforming it gradually toward linearity.

Another new element appearing in human thinking was the dualism, where linearity transforms the continuum of the spectrum into a polarized "image," where are present all the opposing "qualities."

Before the development of the approach toward polarization, everything was about changes and the change was the process that switched of the quality—however, the changes were seen as cyclic phenomena.

Based on the same assumption Dr. Mithen concluded that:

> It was the dualism between hunter-gatherers and domestication that shows that there was, in fact, a continuum of relationship. As the relationship developed between humans, it was also reflected in human relationships with the environment, leading to domestication.
>
> But this new way of thinking that developed the human society and extended this society into the environment occurred only in the last 40,000 years. The seeds of domestication were first laid in mind in the Middle/Upper Paleolithic transition. It is the key epoch that lies at the root of the modern world.

The development of language that took a fast path in the same epoch mentioned above was a practice in "imagination." People begin to think in the way they talk, and the "imagination" used in talking was transferred to thinking. The scope of the language was to control, so the scope of "imagination" was to control, too.

All these interlinked processes were developed by mutations which allowed a lowering of the working entropy. This lowering was also a lowering in energy and information content.

However, this lowering enlarged the domain of the conscious state and also allowed some degraded unconscious information to reach the conscious state.

The language was the voice created by the conscious state while it represented a steady branching out from an unconscious state. This branching or extraction brought language and imagination to the conscious state.

Imagination is an essential component of creativity, but also a capacity that is connected to the memory, problem-solving, counterfactual reasoning, and many other aspects of cognition.

The issue, Psychology, explains,

> some of the most compelling examples are found in "pretend play," a widespread activity in human children that is almost absent in other species. More recently, theorists such as Dorothy and Jerome Singer, Inge Bretherton, and Paul Haris have proposed more positive view linking pretend to play with children's understanding of reality, emotional mastery, and adult imagination. In pretend play, children explore emotions in games that are under their control. They learn *to make sense of the world by considering alternatives to reality.* According to Harris, the capacity to consider what might have happened is crucial to causal and moral reasoning. Children who participated in elaborated pretending will score higher on tests of language ability, self-control, divergent thinking, perspective taking, and a

> large other measures of social and emotional development.
>
> Imaginary thinking continues with an increasing ability to conceptualize nonfactual, fictional, and metaphysical possibilities. It is a uniquely human capacity that develops early and is crucial in everyday thinking throughout life.

In his article, *New Magical Thinking,* Dr. Serge Moscovici of Ecole des Haute Etudes en Sciences Sociales in France, explains his views in the Public Understanding of Science, 2014, Vol. 23 (7) 759–779 :

> Our mind of mind of intellect is relatively new to "modern" man, beginning its evolution again after 5,000 years of "fantasy belief indoctrination." Today we still operate on the ancient wisdom of the past before the invention of the "afterlife scam" and the advent of religions. With our intelligence, we build from "imagination" everything we know of it. But few people are comfortable in their discerning "seeing" mind of intellect. Our intellect is still unfamiliar to us. Most people do not view intellectual thinking as fun or enjoyable. The great search for Truth is still a mystery for humankind.
>
> Children first learn to realize their imaginations, and after their immature child fantasies are dashed, they are thought how to develop their intellectual minds, they are thought to accept adult fantasy as a "new reality" to replace their childhood fantasies.

> Imagination first rejects the thinking of the rational mind in defense of itself. Imagination wants to be free and not constrained. The intellect must invoke imagination... not the other way around. If not...
>
> imagination will run away with you...and away from intellect. Handed down imagination is not working well for society and has not for thousands of years. Protecting belief paramount to a believer and choosing fantasy over reality a badge of honor.
>
> The mind of man has not learned how to develop his logical, intellectual mindfully and viewed logic as imaginary and imagination as reality; the realities are reversing, causing great frustration and confusion.
>
> The imaginary mind will not accept the input of the intellectual mind and in fact protects against it. The intellectual mind gladly draws from imagination; imaginary thinking is paramount to the advancement of any kind.
>
> Magic is the hope to find in it a marvelous secret that normally escapes but deems important to know.

There is a devotion of some people to explore the limits of ordinary thought. The magical has not disappeared, but it just evolved from its ancient traditions to a hard to detect modernity. Is the Law of Attraction one of this evolved kind?

However, everything debated above resulted from a change in the working entropy of the human brain.

EMOTIONS AND COGNITION

In an article of Valentin Afraimovich, Todd Young, Mehmet K. Muezzinoglu, and Mihail I. Rabinovich, published in Bull Math Biol. 2011 Febr (pages 266-284), under the title, *Nonlinear Dynamics of Emotion-Cognition Interaction: When Emotion Does Not Destroy Cognition?* The authors make a novel analysis of this activity in the human brain.

> Emotions and cognition are essential to how we, as humans, respond to changes in our environment. Recent studies in cognitive science suggest that emotion and cognition are subserved by different, although heavily integrated, neural systems.
>
> We study conditions under which the Stable Heteroclinic Channels (SHCs) about cognition will continue to function in the presence of interfering activity with other parts of the network, i.e., emotions.
>
> There are three main mechanisms of cognitive-emotional interaction: involvement of multiple brain centers, like amygdala and prefrontal cortex, in both emotional and cognitive networks; the high degree of connectivity between different brain areas; and the rhythmic brain activity at certain frequencies that supports

> coherent interactions between anatomically distinct regions of the brain during cognitive tasks. An intriguing aspect of emotion-cognition interaction is cognitive control via emotions. On the neurobiological level, cognitive control may be attributed to prefrontal activity inhibiting relevant subcortical emotion processing regions.
>
> Despite the strong coupling between the cognition and emotions, their dependence on time can be quite different. The main difference is following: the emotion may be quasi-static or recurrent in time, whereas a cognitive activity, by the nature of its task, must be transient in time until the termination of the executed cognitive function.
>
> Metastability is a general nonlinear dynamics concept, which describes states of delicate equilibrium. A system is in the vicinity of such an equilibrium, a state where the system spends an extended (but finite) period of time. Under the action of perturbations or interaction, the system is susceptible to fall into another state. However, the fundamental mechanism in carrying out the means of competition is inhibition, which is widespread in the cortex.

The authors of the study conclude:

> It is clear from imagining experiments that emotion and cognition can be represented as the superposition of coarse-grain modes, whose activities are changing with time. The cognitive

> and emotional modes may have distinct roles in different aspects of mental activity. The vibrating part of emotion, unless it is too large in magnitude, facilitates the execution of the cognitive problem as it accelerates the "stream of thought" along the SHC. Too strong emotion pulsation would move the cognitive system out of SHC (or metastability).

In reality, each exit direction would be assigned a certain probability for being followed by the system. It seems that the boundaries of the instability in the emotion system follow a prescribed cognitive function.

This study suggests that each emotion plays a certain cognitive function. As it is known, the emotions are the result of experience, such an experience subserves cognition using emotion.

There is also known that emotions have an unconscious origin. Thus, the unconscious leads to cognition as much as the conscious does.

DEALING WITH UNCERTAINTY

Achim Peters, Bruce S. McEwen, and Karl Friston in their paper, *Uncertainty and Stress: Why It Causes Diseases and How It is Mastered by the Brain,* published in Progress in Neurobiology, Volume 156, September 2017, pages 164–188, explain their findings:

> The ‚free energy principle‘ rests upon the fact that self-organizing biological agents resist a tendency to disorder and must, therefore, minimize the entropy of their sensory states. Applied to our everyday life, this means that we feel uncertain when we anticipate that outcome will turn out to be something other than expected-and that we are unable to avoid surprise. As all cognitive systems strive to reduce their uncertainty about future outcomes, they face a critical constraint: *reducing uncertainty requires cerebral energy.* The notion of the selfish brain captures the character of the vertebrate brain to prioritize its high energy‘. Accordingly, in times of uncertainty, the 'selfish brain' demands extra energy from the body. If despite all this, the brain cannot reduce uncertainty, a persistent cerebral energy crisis may develop, burdening the individual by, allostatic load‘ that contributes to systemic and brain malfunction

(impaired memory, atherogenesis, diabetes, and subsequent cardio- and cerebrovascular events).

Stress arises in those people who are uncertain about the answer. Thus, stressed individuals lack control.

How does the brain make predictions about the world when it only has access to small fragments of it? Apparently, the brain makes use of sophisticated statistical methods. By operating with probabilities, the brain is able to make predictions of future outcomes even under conditions of uncertainty. How exactly these procedures are executed is a fascinating question.

According to that concept, all cognitive and biological systems are driven to minimize an information-theoretic quantity known as ‚free energy'. More precisely, this quantity is ‚variational free energy' and differs from ‚thermodynamic free energy,' though the two are mathematically equivalent. Free energy, defined in the current article as the information-theoretic quantity, bounds surprise, and is conceived as the difference between an organism's predictions about its sensory inputs and the sensations it actually encounters.

Put it concisely, reducing ‚free energy' inevitably reduces ‚surprise'-as measured by a violation of predictions. In exactly the same vein, reducing, expected free energy' inevitably

> reduces, expected surprise'-known as entropy or uncertainty. Thus, in the long-term, we are all compelled to avoid surprises and resolve uncertainty.

The brain needs the energy to process the information that bears the uncertainty (that information is here "free energy"). *To be able to get more energy, the brain must erase some of the existing information that is "free energy."*

Here, energy is traded for energy.

However, the paper I referred to in this chapter suggests that humans cannot deal properly with uncertainty because of an energetic constraint. This idea can explain why a larger type of human skull (and it is assumed a larger brain inside the skull) dominated for almost eighty thousand years but made a U-turn since twenty to ten thousand years ago? Why did that type of skull encountered a significant diminishing in size in recent times?

It may be so because the Late Paleolithic brain was dealing with more uncertainties, especially after the venture in the out-of-Africa migration (that refers again to the last eighty thousand years). The process of deglaciation, the introduction of domestication, the establishing of semi-permanent and permanent settlements and the technological progress were all elements that contributed to reducing the overall uncertainty of the ambient.

To the reduction in the overall ambient uncertainty, it could be added the change in diet caused by domestication. This diet produced less energy. Consequently, the brain had to find a way to reduce its dealing with uncertainty by selecting information that presents lesser variability (uncertainty). This new way favored the adoption of linearity.

What about the present time when we discovered that uncertainty is still prevalent around us? Probably, we must have in mind a set of answers without necessarily knowing which is correct.

However, in sum, dealing with uncertainty requires a lot of energy and erasing some existing information does not suffice to accommodate our attempt to compute nonlinearities, which are uncertainties. Humanity must find a way to provide to our brains such extra energy.

ESSENTIAL REMARKS

1. Solovki and other geomagnetic legacies
2. Consciousness evolved to facilitate communication but instead introduced programming and control
3. The dangerous fascination of imagination
4. Virtual reality reveals important issues about the brain mechanisms

1. Solovki Legacy

The geophysicists agree that several important geomagnetic excursions occurred since the current interglacial begun. They are: Hilina Pali (18,000-17,000 years ago), Gothenburg (13,000-12,000 years ago), a no-name excursion (11,000-10,000 years ago), Solovki (7,500-4,500 years ago), and Sterno-Etrussia (3,000-2,300 years ago).

As one can see in the above paragraph—the longest and hence, the most significant geomagnetic excursion was Solovki event (5500 BCE-2500 BCE). It historically corresponds with the first major steps taken in the evolution of our civilization because Solovki event extended on three thousand years.

Many details about this geomagnetic even were previously given somewhere else in this text.

As noted before, the geomagnetic events, especially the long ones, are prone to produce a chain of small mutations in biological systems. In the case of humans, the setting on of the civilization implies the spread of agriculture, first settlement as villages and cities, megalithic structure, monuments, etc. But all these achievements were intermediated by a sudden manifestation of strong imaginary thinking.

As stated before, all genetic changes represented various alterations of the working entropy in the human brain.

As it seems, the cosmogenic radiation, penetrating to the ground, during the geomagnetic excursions, produced local entropy increases in human brain. For not very clear reasons, these increases affected in human brain the entropy of the frontal lobes. As an adaptation to entropy increasing, the frontal lobes developed the strongest system of suppression (the called filters). This advanced system of suppression was part of the genetic mutations occurring in the brain.

There is to assume that because of the higher suppression system developed in the frontal lobes, other regions of the brain ceded energy (entropy) to frontal lobes because they can suppress; this process was allegedly contributing to the asymmetric development of the lobes.

The entire process of increasing the entropy and the consequent suppression favored the development of language mechanism, but also advantaged the extension of the conscious in the same region of the lobes.

The study and scanning of the brain under meditation indicate the process of accumulation of entropy and respective suppression in the frontal lobes, turning other sections of the brain to dormant states. The meditation reduces the entropy of the frontal lobes and helps spread the energy toward the so-called dormant regions, increasing their local activity. However, these dormant regions are considered to be more primitive.

As a result of the geomagnetic stress, which became very significant since forty-two thousand years ago (starting with Laschamp Excursion and restarting anew thirteen thousand years ago with Gothenburg Excursion, but the main culprit for mutations was Solovki Excursion), the human brain was forced to accept the impact of the incoming cosmogenic energy. It operated several local adaptations, like the redirecting of the excessively gained energy by most neuronal systems toward the frontal lobes and local higher development of suppression in the same area of the frontal lobes. All these recent processes created our "modern brain."

Undoubtedly, the most revolutionary aspect of the "modern brain" became the "imaginary thinking" that was intermediated by the development of language.

* * *

The epoch of Solovki's geomagnetic distress (5500BCE-2500BCE) corresponds with several large genetic adjustments, where one can observe that the newly formed "imaginary thinking" have turned into practice.

The haplogroup R1a (Aryans), living together with their minoritary branch of R1b, is found extensively in Eurasia where this genetic conglomerate gave rise to several major cultures, like Combed Ware Culture (5500 BCE-2600 BCE) that was absorbed by the Corded Ware Culture (3200 BCE-2350 BCE). At the time when Corder Ware Culture was formed all over northeastern Europe up to Vistula River, another important culture was formed to its east: the Yamna Culture (3300 BCE-2600 BCE) that extended to the Urals. In the same period, and simultaneously, another culture was formed from the Urals to Lake Baikal while allegedly reaching into the Siberian Far East; this was the Afanasievo Culture (3300 BCE-2500 BCE).

It is interesting that the DNA analysis found that all these three cultures belong to a single haplogroup R1a that lived together with the minoritary R1b haplogroup.

In other words, for one thousand years (around 3500 BCE to 2500 BCE) from East Germany to Scandinavia and the Pacific, in all northern and central Eurasia, was basically one single but multi-tribal culture named the Aryan Culture. It was like the largest empire that ever existed

Later, in the third millennium BCE, it started that multi-tribal Aryan conquest of the rest of the Eurasian world that evolved in three consecutive migration waves and reached Greece, Middle East, India, Egypt, and toward the east the Aryans reached all of Siberia, Mongolia, and also the northern and western parts of China.

However, at the start of the geomagnetic distress, some 5500 BCE, also originated two important cultures in Balkans and Eastern Europe like Starcevo-Vinca-Cris interlinked cultures, and the Cucuteni-Tryppilian culture. Both populations had a preponderant Cro-Magnon DNA.

In the Middle East, around 5500 BCE, the Ubaid Culture appeared and was followed around 4000 BCE by the Sumerian Civilization.

The archaeologists agree that the Indus Valley Civilization started around 5500 BCE. They also agree that the founder of the Chinese Civilization, the mythic Yellow Emperor could be dated 2700-2600 BCE.

In Egypt, the predynastic cultures started 5500 BCE, and unified Egypt appeared in 3150 BCE. The Old Kingdom was from 2700 to 2200 BCE.

Is this such a type of coincidence that the civilization took off around 5500 BCE, and this corresponds with the beginning of the geomagnetic distress that lasted for the next three thousand years (5500 BCE-2500 BCE)?

Here, must be said that an important requirement for the beginning of the civilization was the social element and its capacity to communicate effective, efficient, and extensively.

This communication was the basis to produce and spread imaginary thinking.

The imaginary thinking created the civilization.

2. Consciousness Evolved to Facilitate Communication.

Roy F. Baumeister and John A. Bargh in their publication, *Conscious and Unconscious-Toward an Integrative Understanding of Human Mental Life and Action* (2014), came to a very important conclusion: "*conscious thinking is for talking.*"

> One crucial function of conscious human thought would be for communicating. Although many things can be done while conscious thought is directed elsewhere, talking does not appear to be one of them (nor does writing).
>
> The assumption that communication is one central purpose of conscious thought offers a useful solution to several dilemmas. We both think automatic and unconscious processes do a fine job of executing all behavior. There is no evidence that conscious can dispense with unconscious processes in action.
>
> The adaptive benefits of communication seem uncontroversial. Sharing information is something humans do extensively, and more extensively and deliberately, far more frequently and more effectively than any other species. Natural selection might relatively quickly have favored communicative abilities.
>
> Logical reasoning seems to require conscious thought (though, as always, in conjunction with automatic and unconscious processes). The presumptive advantage of reasoning would

seemingly constitute a possible explanation for why consciousness evolved. With reasoning, too, however, there is a strong case that the driving force has been the interpersonal communication and interaction rather than solitary or solipsistic ratiocination. Reasoning evolved primarily for the sake of arguing, in which case many of those seeming flaws appear instead as helpful to the cause of survival. Logical reasoning and planning can also be helpful in guiding behavior toward optimal, desired outcomes, and these appear to require conscious thought.

Impulses to action arise unconsciously and are then interpreted consciously to form a coherent narrative account of what one is doing and why.

The unconscious evolved primarily to produce adaptive responses in the immediate present. Conscious processes evolved to do other things. They are necessary for people to time-travel away from the immediate present, especially to anticipate, simulate, and plan for the future. They are necessary to communicate verbally with others. One prime adaptive advantage of being able to communicate with others seems to have been the production of good arguments, both to persuade others to help one pursue one's own important goals, and especially to defend one's sometimes deleterious behaviors to one's social group, to avoid retribution and ostracism.

Insofar as consciousness evolved to facilitate communication, and human communication continues to take new forms and present new challenges and opportunities, *it is tempting to speculate that human consciousness itself may continue to evolve.*

3. The Dangerous Fascination With Imagination.

As Christopher Gergland explains in his article, *Imagination Can Change Perception of Reality,* published on June 28, 2013 in *Psychology Today*:

> Researchers in Sweden have found that our imagination can change our perception of reality. Your mind can play tricks on you by changing illusions of what you think you hear and see into what seems like reality. The new study from Karolinska Institutet is published in the scientific journal Current Biology.
>
> We often think about the things we imagine and the things we perceive as being clearly dissociable," says Christopher Berger, a doctoral student at the Department of Neuroscience. "However, what this study shows is that our imagination of a sound or a shape changes how we perceive the world around us in the same way actually hearing that sound or seeing that shape does. Specifically, we found that what we imagine hearing can change what we actually see, and what we imagine seeing can change what we actually hear.

According to Professor Henrik Ehrsson, the principal investigator of this study:

> This is the first set of experiments to definitively establish that the sensory signals generated by one's imagination are strong enough to change one's real-world perception of a different sensory modality. According to the scientists at the Karolinska Institutet, the results of the current study may also be helpful for understanding the mechanisms which the brain fails to distinguish between thought and reality in several psychiatric disorders.

In an article by Christopher Wanjek titled *Imagination and Reality Look Different in the Brain* published in *LiveScience* on December 23, 2014, the author explains:

"There seems to be a lot in our brains and animal brains that is directional-that neural signals move in a particular direction, then stop and start somewhere else," said Dr. Giulio Tononi, a psychiatry professor and neuroscientist at the University of Wisconsin-Madison and his collaborator Daniela Dentico. The results of this study were published in November 2014 in the journal NeuroImage.

> A leading theory in image processing posits that our visual mental images are not stored somewhere in the brain, but get actively reconstructed. The brain does this by reversing the order it uses for visual perception, showing the prevalence of top-down signal flow during imagery.
>
> The researchers could not determine, however, whether imagination originates in the parietal lobe. It may instead flow through the parietal

> lobe from the frontal lobe, the brain region most associated with human intelligence.

Dr. Michael Kalpakgian in an article published in Seton Magazine on August 5, 2014, expresses his opinion about Dangerous Dreams:

> The dangers of imagination prevail when man ascribe to himself greater importance, knowledge, or power that he actually possesses, when man is deluded into thinking that heavenly happiness is possible, when man loses contact with reality and dwells in the mental universe of ideas, abstractions, and theories by inhibiting ivory towers, living solitary lives, and inventing false views of the future that have no correspondence with the wisdom of the past. The dangerous imagination ignores the structure of reality and the nature of things. How often the mind, hurried by her own ardour to distant views, neglects the truths that lie open before her-truths that lie open to common sense.

I will end these excerpts by quoting Carlo Rovelli, a theoretical physicist at Aix-Marseille University, France.

> A number of my colleagues in theoretical physics have spent their life studying a possible symmetry of nature called "supersymmetry." Experiments in laboratories like Geneva's CERN seem now to be pointing more toward the absence that the presence of this symmetry. Some colleagues said: How dare Nature not conform to our imagination? In theoretical physics, technical journals are replete with multi-universes, parallel

> dimensions, scores of particles that nobody has ever seen, and so on.

To a larger extent, we live in narrations we weave ourselves. We can create enchanting explanations, images of ourselves, of our own great country, of our great society. We can be fascinated by our own dreams.

> *But something tells me that we should worry. We live inside a real world, where not all the stories are equally good, equally effective. One dream out of many is the good one. Few explanations are the correct ones. More often than not, our wild imagination turns out to write poor drafts with respect to the surprising variety of reality. I worry that we are going overboard with our contemporary fascination with imagination, and, in so doing, we risk losing track of the harsh independence of the world from the weakness of our mind.*

Professor Daniel Dor of Tel Aviv University wrote an article titled, *The Birth of Human Imagination Out of the Spirit of Language*, published in *The Junkyard* on April 4, 2018.

The author said:

> Language is collectively-constructed and cognitively-sustained communication technology.
>
> Language is the only system that goes beyond the experiential strategy. It works with what I call the *instructive* strategy. It allows speakers to intentionally and systematically instruct their interlocutors in the process of *imagining*

the intended experience-instead of directly experiencing it. The speaker provides the receiver with a code, a plan, a skeletal list of the basic coordinates of the experience-which the receiver is then expected to use as a scaffold for experiencial imagination. In instructive communication, the sender communicates: my experience is of *this type*-try to *imagine*.

Language is thus the only system that allows for communication that actually bridges the *experiential gaps* between speakers-*through the systematic instruction of imagination*. The more your words force me to imagine-the more communicatively valuable they would be for me.

What is unique about human imagination is exactly the fact that it is creative: we can re-arrange and intersect our past experiences that go far beyond what we have actually experienced.

Every advancement in the technology required more imagination; every new imagined experience required further technological development. The more dependent human society becomes on the technology of language, the more individuals in it were selected for their imaginative capacities. **With time, society itself came to acquire imagined properties in and of itself**.

To summarize: *all linguistic communication requires an effort of imagination on the side of the interlocutors.*

4. Virtual Reality Reveals Important Issues About Brain Mechanisms.

The publication, GIZMODO, has on March 18, 2015, an article by Sarah Zhang titled *The Neuroscience of Why Virtual Reality Still Sucks*. This article incorporated several interesting facts and made suggest some additional explanations. Hence, I quote from it.

> True, the past year has brought a great flourishing of virtual reality systems that are miles better than the clunky, nauseating devices of the 90s. This new set of devices is good enough to feel stomach-dropping real-even though the images are still pixelated and lag a tiny bit. When journalists write about being wowed by the latest fancy VR device, they don't mean that VR is indistinguishable from reality. As Jason Jerald, a technology consultant for VR companies puts it, "We can get very engaged in cartoon-like words." Images don't have to look perfect for presence. But these imperfections become obvious if you spend more than typical few minutes of a press junket inside VR. There are many reasons, both conscious and unconscious, that your brain rejects the reality of a screen mounted a few inches in front of your eyeballs. The main reason is latency or the tiny but perceptible delay between when you move your head in VR and when the image in front of your eyes changes-creating a mismatch between the motion we feel (with our inner ears) and the image we see (with our eyes). In real life, this delay is essentially zero. "Our sensory system and motor systems are very

> tightly coupled," says Beau Cronin, who earned his Ph.D. in computational neuroscience at MIT.
>
> A weird thing happens in VR: You can look at the far-off horizon of a virtual beach but still feel like you're in a room. At its worst, it can cause painful eyestrain and headaches. The objects rendered through light are supposed to achieve true depth, coming in and out of focus as a real object might.

"The larger the field of view the more sensitive you are to motion," says Frank Steinicke, a professor at the University of Hamburg. That's because your peripheral vision is especially sensitive to movement. Peripheral vision appears to be closely integrated with your sense of spatial orientation, being separated in the brain from the pathway of central vision."

Now, the VR experiments show that, when being in VR mode, 60 percent of the neurons stop working; the same VR mode fails to make a space-map in the brain.

In my opinion, the fact that the VR device has its screen three to four inches away from your eyes makes the brain to adopt a reading-like process that is language related and uses verbal circuits. It eventually suggests that only 40 percent of the neurons will work in verbal circuits, while the rest of them are still dedicated in exclusivity to pure vision. For the same reason, the verbal circuits fail to make a space map that is habitual only for vision through some depth.

Another argument to complete the above is the fact that language was developed as a near-field aspect of reality. As far as the VR devices are currently constructed, they exploit a near-field technology that will always stimulate the verbal circuits.

However, the experiments indicated that motion is sensed not only by vision but also by all other senses. The result is a complex feeling where all senses participate. This result is similar in conclusion with a several years ago a study of MIT that disclosed that "understanding" given by the work of all senses combined is more important than "problem-solving" ability.

Here, I have to say that, in my opinion, VR is related to "imaginary thinking" in the sense that both are based on "synthetic reality," and both are intimately connected to the evolution of language and "verbal thinking."

UNIVERSAL TECHNOLOGY OF LANGUAGE

One may consider the language as a tool or a toolset. But such toolset assembles things similarly with every known type of technology.

Hence, it is a technology in itself that can produce an infinity of other technologies by playing with its assembling/disassembling capacity.

All technologies we ever produced must follow up this mindset that manipulates and assembles. For this sake, all technologies are all but a disguised copy of our language, or any of our languages because verbal and its utterance are not the only languages: here, there are the sign, visual, musical, mathematical, and much more other languages. All these languages use the same process of assembling and disassembling, known more generally as analysis and synthesis.

As Professor Dr. Jon Dron with Athabasca University, Edmonton, Canada, has stated in a seminal paper with the title "Is language a technology?":

"We can use language to manipulate ideas, create and transform concepts, design, explore, analyze, and more to achieve some goal or goals. We can use language to manipulate language, to construct things in language, and use those constructions to make other constructions".

"It is in many ways the fundamental human invention, hugely more important than everything else, like fire, the wheel, or the

Internet. To be human without language is barely conceivable; a lack of language to become widespread, we will no longer be humans".

"Language is the ability that enables us to symbolize, abstract, construct, amplify, and enhance, but also it generates the opportunity to spread technologies, build ideas, learn, create, discover."

"It's a wonderful virtuous circle that leads to an ever-expanding explosion of knowledge in our species as a whole even though we, as individuals, are likely getting dumber and are very likely dumber than some of our distant extinct cousins. It is not intelligence that makes so successful as a species: it is how we use technology to amplify that intelligence".

Assembling put together the combined properties of the elements. Combining several simple toots, we create a complexity of them, where the original properties turned significantly changed, while some properties become amplified.

Humans before language were very similar to all other animals. These other species could climb a few steps up on the scale of intelligent evolution. But language made us climb to the top, and continues to amplify such a distinction.

I would say that the "intelligence" is in the language, and it is transmitted to the technology. In other words, "intelligence" is the result of "assembling." Intelligence naturally resides in "complexities" that result from "assembling."

As some scientists see it, we have turned dumber in the last century, while our power of assembling increased; even then, we are capable to "assemble" an ever-increasing number of new technologies.

Why occurs such an amplifying process that, once started, could not be stopped except by a self-induced extinction?

The language enables us to find a structure for thinking in the utterance. It means the language transforms unstructured data into structures, which become available to be maneuvered by cognition by "assembling" and "reassembling" them.

It has been argued that "language is natural collective technology that evolved primarily to facilitate efficient communication in populations" (Salikoko S. Mufwene, University of Chicago).

It was found that language implies several other things, which are as important as the communication itself: they are planning (all utterance is planned in mind), control (the reason of language is to control the audience, and by such control, the utterance can be shared), socialization (there is no reason one to talk for himself only).

Darwin was positive that language was a prerequisite for "long trains of thoughts." But such trains of thought have no meaning at all if they cannot be controlled. On the other hand, the "train of thoughts" is unstructured data. By controlling such data, we structure it.

Thus, in my opinion, the most important advent brought by language was a new capacity to structure the unstructured information and data.

CONCLUSION

In short, our civilization evolved by an increased capacity to discover new options and managed to manifest them in physical products. This fundamental but new capacity was provided to us by the *evolution of language and imaginary thinking*.

In my opinion, several changes in gene expression and mutations affected the majority of planetary populations. Every mutation must create a competitive advantage. A fast changing environment must force the organisms to produce multiple random but small mutations, which are the only ones that fit to surviving (there is known that large mutations are dangerous, and might lead to the species sudden extinction).

Each genetic change in the human brain has opened a new window into the unconscious. Consecutive genetic changes opened multiple windows of opportunities. Probability is defined as a ration between outcomes and the number of trials. Here, each mutation has a probability of uncertainty: it may succeed or not.

Transliminality reflects the threshold at which unconscious processes enter into consciousness. Some individuals have a predisposition toward transliminality, and the neuronal studies indicate that different brain rhythms have a distinct activity in such predisposed brains. Also, the research found particular altered hemispheric asymmetries in the human brain, which are developed in the predisposed brains. Hence, it is a neural correlate of transliminality.

However, transliminality defines those individuals who claim to experience paranormal phenomena. The reason that most scientists dismiss paranormal claims and their research is that no one has any idea or suggestion about how the brain could interact with the physical world.

Even then, along with ancient history, there is to assume that some significant achievements were based on a sort of paranormal effects.

In the end, the paranormal seem to contain and deal with information that exists beyond the one-in-a-million conscious selection of reality. As the research shows, such information populates different areas of the brain that are normally dormant, and uncommon neural connections and brain rhythms excite that information content.

* * *

The conclusion must deal with two distinct aspects:

- the brain capability to extend its reaches into the analysis of the perceived information, where today ration is as poor as one-in-a-million;
- the second aspect is related to the fitting of our artificial world shaped by imaginary thinking into the natural world, and any conflict of interest that might arise.

The first part might deal with the theoretical possibility to generate a scale-independent property of a unified unconscious hierarchy of scales that is called hyperscale. There are several questions to address:

- Is the goal of cognitive control to produce a total elimination of "uncertainty?"
- Is "uncertainty" inherent in nature?

If the answer to the second question is positive, this means that we can never get rid of "uncertainty."

And then, comes another question: how close to a manageable outcome can be maneuvered by this "uncertainty?" How much truth does this manageable outcome contain? Would this be no more than a negligible quantity like 1 percent or less?

Still, another question might arise: is any other way we can use to "learn" or "know" much-of-the-truth without directly involving the "uncertainties" and "probabilities?" Like here, can we compress time so much as to get into the deterministic order that rules the chaos? Or would we be able to find a way to understand and deal with natural hierarchies?

In my opinion, the most important issue to be addressed is about emotions (which result from actions) and feelings (which come out of our perspective on those actions), because both allow us to have "creations" in response to them.

We would not have music, art, religion, technology, economics, politics, justice, or moral philosophy without the impelling force of sentiments and feelings.

In standard usage, the term, "emotion," is used to designate "a state of consciousness having to do with the arousal of feelings. (Webster Dictionary).

Now, "emotional intelligence can reasonably be conceived as a measure of the degree to which a person successfully or unsuccessfully applies sound judgment and reasoning to situations in the process of determining emotional or feeling responses to those situations. It would entail, then, the bringing of (cognitive) intelligence to bear upon emotions. Critical thinking provides the crucial link between intelligence and emotions in an emotionally intelligent person."*

---

* These quotes are from Linda Elder from her paper Critical Thinking and Emotional Intelligence.

In more recent research, there is shown the conscious flow is easy to change, but the unconscious flow is highly resistant to conscious attempts to change it. Unconscious mind usually exerts priority in emotional response.

Undoubtedly, emotional thinking had an immense role in creating human cultures.

But here, the "imaginary thinking," being developed starting with 5th millennium BCE, and becoming theorized in the 5–7 Centuries BCE in Ancient Greece has been the one that completely changed the course of human history while ultimately leading to the exponential development in all technological fields.

However, the answer to those questions is beyond the scope of this paper.

Let's see about the second aspect.

The overlapping of mandatory education, the introduction of the math as another language, and the development of artificial computation, all contributed to making the unconscious windows increasingly complex. Today, we benefit vastly from such complexity which characterizes our current unconscious selectivity.

We currently deal with our production of information which doubles every year and which tends to challenge natural information that is selected by our oculomotor system. From 11 million bits of information that enter the oculomotor system, the processes which deal with attention can select 30-60 bits with each glimpse. Concepts of uncertainty and redundancy are closely related to attention.

Every single moment, our oculomotor system discards millions or possibly billions of bits of natural information. A similar case seems to occur in selecting the information produced by our civilization. We have created the possibility that this artificial information of ours can be stored in a type of artificial memory.

We try to replicate what our brain and unconscious selectivity do artificially. Even if we succeed, this will not solve anything regarding our brain processes. We just transformed everything into an anatomically-based civilization of mentally generated multiple options. While these are naturally provided options, their implementation created an artificial world that is far less complex than the nature it tries to replace.

The entropy of the visual stream reaching the human eye is about 10 billion bits/second, while only 3-11 million bits/second leaves the retina for the brain systems, and less than 10,000 bits/second come under attention scrutiny but only 30-60 bits per glimpse. It means that human attention deals annually with 31.5 billion bits and in the meantime, the *brain systems annually receive 10 trillion bits of information*.

Our attention, to be able to process the incoming flux of information, reduces by one million times the brain entropy.

By contrast, our civilization has produced in 2016, 2.5 quintillion bits of data.

However, this amount of data doubled in 2017. What is the impact of all such data in our brain?

Every biological system is driven by agents acting individually, asynchronously and randomly; against this apparent disordered activity, the biological systems accomplish complex and very precise tasks. This biological computation is significantly different from our mental or artificial computation. A non-analytical natural computation produces these biological results. Similar situations and results characterize our entire brain activity.

However, all biological systems, our planetary systems, our planet as a whole, all of them, produce natural information, and all such information is processed in one way or another by various natural abilities.

Here, we have to consider that some information entering our brain is naturally produced but some other information is not natural, being increasingly provided by our civilization. During the last several decades, the rate of growth of artificial data generation has reached exponential figures.

The question is: How does the brain process natural and artificial information? Does the brain compute the natural compared with artificial information in the same manner? How distinct is natural computation when compared to our mentally inflicted type of computation? Does cognition involve information processing? Could non-natural information be assimilated or reduced to natural information? How distinct is non-natural information compared to natural information? Is information the basic element of our universe?

These questions are very important but very difficult to answer for the time being; as it seems, they would remain open questions.

Natural information is carried by neural and cognitive events in correlation with environmental elements. But here, natural environments increasingly transformed by our civilization, carrying a decreasing number of natural elements; thus, they carry a rapidly increasing number of artificial exhibits.

Maybe cognition in itself does not involve computation or natural information.

The artificiality we have created does not help us to understand much of the natural world. Why? Because whatever we have created remains alien to nature and its mechanisms. It is a type of artificial invention that is built out of anatomical mechanisms and constraints.

Our base assumptions originate in the information selected by the conscious states of the mind that are a reduction, by at least one million times, of the surrounding reality. This type of selection helps us to survive, but this is of very little use in understanding reality as a whole and its true mechanisms.

There is evidence that consciousness, everywhere it exists (and humans are not an exception but the norm), selects from an immense amount of information a tiny fraction that is used to produce that part of the interacting universe we perceive. Such a selection varies from one entity to another. The unconscious is variously stratified and variously related to the selection made by consciousness.

Humans have been able to reconnect the conscious and some of the unconscious selections in a novel way that creates a non-natural mechanism. Was this an accident of evolution? How frequently do such accidents occur in the universe? Is there a universal faith for all such accidents? If the universe is a computer generation, then something or someone created that natural computer. In this particular case, how is our accidental and artificial evolution perceived by that natural computer? Did we produce a conflict of interest, or not so?

* * *

Here in every moment, our civilization is confronted directly with the immensity of data added in short-term to an even larger immensity of data that already was created.

The data is information, and this information contains an amount of "uncertainty." Our consciousness selects only the information showing less "uncertainty," but the quantity of such type of data rises exponentially. And then, the tiny amount of "uncertainty," being maintained in each selected information, adds exponentially, and globally the accumulation of "uncertainty" increases immensely. Somehow, we deny this global reality because we think that each piece of information in use has, by its original conscious selection, only "certainty."

Now, in every split second, this extreme issue steadily grows, leading our civilization to an immense global "uncertainty."

Many specialists see "uncertainty" as entropy. Rising "uncertainty" increases the entropy of our civilization globally, and here, there are no built-up filters to suppress the entropy. Is our consciously created "world of order" in danger?

However, other specialists see "uncertainty" as stress. Hence, stress grows exponentially, and it affects the human biological system starting with the brain and ending with whatever the brain regulates in our body.

Thus, stress becomes the main enemy of our civilization. Stress generates illnesses, and such "illnesses" grow exponentially, too.

In sum, the information generated by our civilization equals the same amount of stress. The original intent of the civilization, being created by our "imaginary thinking," was to eliminate as much as possible of the surrounding "uncertainty," but it ended up by producing an immensity of information, "uncertainty," entropy which put in severe danger human mental health and body systems.

The estimates show that around 2045–2050, human society can be affected one hundred percent by at least two major illnesses like the Alzheimer and autism, while a whole bunch of other chronic illnesses will overwhelmingly drag on the side. Are we approaching an extinction?

PART TWO

JUSTIFYING THE HYPOTHESIS OF HOMO LOQUENS

WHY THE MODERN BRAIN FAVORED THE LEFT HEMISPHERE?

In the late Paleolithic, about forty thousand years ago, the human brain showed a right hemisphere dominance. It is called the hemisphere of the visual mode that reflects an accumulation of features, which occurred in several million years of natural selection. The basic features of the right hemisphere were *visual thinking* and *emotional thinking*. It was the place where *art and music* became actively expressed when sudden progress of the whole brain was recorded about seventy thousand years ago.

It could be said that the entire brain benefited from the skull enlargement that reached its peak in the era seventy-five to fifteen thousand years ago.

This process of enlargement coincides with preserving the domination of the right hemisphere. Let's see what the right hemisphere have provided?

It provided ***top-down approaches*** derived from vision, which were synthetic in nature, and created "generalization-processing."

Natural selection favored a synthesis of the behaviors manifested by reality. In short, it developed a "correlation-thinking" that dealt with "nonlinear aspects of reality."

As the brain grew in size, all these features and others developed.

As mentioned in other articles of mine, the geomagnetic disturbances (called geomagnetic excursions) allowed an increased almost to double the concentration of C14 isotope present in the terrestrial atmosphere. And C14 high concentration increased the activity of the Dentate Gyrus in the hippocampus, producing an increase of plasticity.

It is known that DG plays a significant role in pattern separation. Hence, simultaneously evolved the pattern separation in heft hemisphere and the features of the right hemisphere. Pattern separation helps analysis, while increased analysis power forces the synthesis to evolve, too. The evolution of synthesis (as a dominant) was gradually losing the balance to newer capacity for analysis until Early Antiquity. Even then, the analysis that was evolving in detriment to synthesis changed the balance, becoming dominant.

The new path was switching from generalized approaches to an approach for detail.

For a while these features evolved as a whole, being the eventual reasons (along with changes in diet) for which the human skull size significantly enlarged, reaching its peak around twenty thousand years ago.

In the epoch seventy thousand to thirty thousand years ago, the progress in language was relatively little. As the research indicates, the revolution in language occurred less than thirty thousand years ago, gradually accelerating toward Antiquity.

The C14 isotope higher concentration was embedded in neuronal DNA and increased to double the brain plasticity. This plasticity manifested through a major increase in the analysis (pattern separation) but also increased the memory storage and memory retrieval capacity.

For a while, it also contributed, but for a smaller extent, to increases in synthesis. All these new features were determinant for language formation and evolution.

Gradually, the analysis was favored against the synthesis. The analysis increased the available details of the reality, making another change to occur: the qualitative thinking was rapidly losing ground against new quantitative thinking. This new element was based on developing superiority of the left hemisphere for topological perception (perception of shapes and geometry). It was so because left hemisphere evolved into a more piecemeal memory of the scene as being focused on salient (important) parts of the scene (constructive role) rather than on the whole thing (holistic view).

The constructive memory role, well represented in the left hemisphere, essentially contributed to the evolution of language, and gradually imposed the dominance of this behavior (constructive behavior) over the entire brain.

The constructive behavior was paramount in creating the new thinking component of "imaginary," as "imaginary thinking."

The language changed everything in the architecture of the human brain.

The language is essentially linear and gradually thinking turned into "*thinking-as-talking*." Hence, the nonlinearity and holism of vision lost ground to "*verbal thinking*." Synthesis lost ground to analysis. Emotional thinking lost its dominance against "*logic thinking*" derived from analysis-synthesis combined features.

Topological perception unveiled the geometry of things. The analysis helped find the relationship between geometrical elements, gradually generating "*mental computation*" and later the "*mathematics*."

Everything originated in language development, and everything consequently evolved in the left hemisphere.

The analysis helped make a distinction between "one-self" and the others that build up the "egos."

All new features added to the left hemisphere were selected as elements of consciousness because they behave with practicality.

The Dentate Gyrus of the hippocampus is placed in both hemispheres, but the effect of the incorporation of higher levels of C14 in the neuronal DNA favored a neurogenesis that mostly affected the left hemisphere.

The same geomagnetic disturbances (as mentioned in other of my articles) also stimulated an "***increase in the brain rhythms***." As it appears, *the most significant increase occurred for theta rhythm that coordinates the development of language*.

The investigation that attempts to answer this question continues in the next chapters.

CONJUGATES SWITCHED THE BRAIN LATERALIZATION AND GENERATED A NEW TYPE OF INTELLIGENCE

A conjugated reaction refers to temporal spikes emitted by neurons as a result of increased energy provided to the brain by an increased entropy.

The research results from various experiments suggest that increased stimulus generates an increase in inhibition.

However, blocking-inhibition in random neural networks reduces response variability and increases the sensitivity of activity within the location of the stimulus.

In this case, the inhibition acts as a dispenser and sharpens the stimulus contrast (Netta Haroush & Shimon Marom, March 19, 2019, in Scientific Reports 9, Article number 4969-2019).

In the situation of ***increased entropy***, *the above-mentioned* ***stimulus acts to disperse the inhibition, while enhancing the stimulus only locally***.

In simple words, the increased entropy is increasingly filtered by the inhibition; as a result, the neural activity enlarges, and an increased amount of information is provided to the conscious domain. It augments the consciousness.

However, the brain hemispheres have developed a type of conjugated activities: the right hemisphere is characterized by hyperactivities, while the left hemisphere shows hypoactivities, which are inhibitory in nature.

For example, the control of the right hand is driven by the left hemisphere; this hand is the most used because the processes of inhibition assure the most accurate control of the right-hand movement. Prehistoric people were hunting and building their tools with an overwhelming contribution of their right hand.

The ***right hemisphere*** locates the visual activity that *implies a lot of* ***sensitivity***. It is known to be the seat for music and rhythmic acoustics. It was the dominant hemisphere that generated the visual-thinking. Thus, its inherent sensitivity allowed the development of emotional thinking.

Not coincidentally, the sign language developed as a first because of the role of the right dominant hemisphere that was guiding the visual element of signing. Hence, signing evolved from visual thinking.

The research has determined that lip-smacking (as a type of sign language), as observed in several primates, occurs in the same theta rhythm range as human speech. So, both sign and vocal languages seem based on a pre-existing perceptual neural oscillation, like the theta rhythm.

However, the control of a more complex signing needed the control and planning of the neural capabilities mainly developed in the left hemisphere.

The evolution of language needed a better and more complex modality to communicate, and as their anatomy evolved, humans opted to refine the acoustic component.

At that time, the sign-language evolved by inserting grammar rules, which rules came from that hemisphere with higher capabilities to control and plan, which was the left hemisphere.

The acoustic component of language needed an even higher type of control (for modulation and grammar), and therefore, the only option available inflicted the full contribution of the left hemisphere.

As the language evolution turned into a success, by better communication, improved planning, and control, an increased rate of socialization and domestication/self-domestication, the role of the left hemisphere became increasingly prominent.

At some point in time (assumable some 4,000 years ago), the hemispheric dominance switched completely, making the left hemisphere the ruler of the brain.

I used this short review of the *visual conjugated relationship with the evolving verbal mode* to suggest that **our brain evolution was based on a conjugated relationship between the brain hemispheres**.

Even when the brain lateralization is similar in all animals, our prehistoric ancestors succeeded to use a narrow niche that contributes to developing a novel type of communication among the mammals. It was the effect produced by the sensitivity to the initial conditions.

The switching in hemispheric dominance seems to be a novelty in the animal world that generated an anatomic type of technology (language). There are known other beings that use an anatomical technology, but the complexity developed in humans exceeds everything that is currently known.

Maybe the most important effect of such switching is reflected in "intelligence." Human visual nonlinear thinking changed into verbal linear thinking, and **right brain intelligence turned into left-brain intelligence**.

Here, I have to say that **these two types of "intelligence" evolved in a conjugated relationship**, while being a direct result of the effect (while temporal) of C14 isotope on the DNA architecture of the Dentate Gyrus in the hippocampus and affected the very process of neurogenesis.

The high concentration of the atmospheric C14 isotope repeatedly occurred in the last 42,000 years, and it repeatedly but temporarily affected the Dentate Gyrus architecture, suggesting several corresponding neural mutations **or significant changes in the neural gene expression**.

In sum, after several transitional stages, starting 42,000 years ago, an ever-accelerating process of dramatical changes occurred between 10,000 years ago and 4,000 years ago. I would presuppose that about 4,000 years ago, the new architecture of the human brain was already established, and the new intelligence became effective.

As Cheri Florance, Ph.D., explains, "the eye-to-brain pathway is called *opticoders* (the visual brain), and the ear-to-brain pathway is called the *lexicoders* (the auditory-verbal brain). Even today, 32% of the population is born with a visual brain. This means that these people use their *opticoders* to think and process information". "The ***opticoders* brain is considered to be 20,000 times faster than the lexicoders brain**, leading to the type of visual, outside-the-box thinking that has resulted in some of the world's greatest inventions and creative contributions".

The above paragraph indicates what the difference between the old visual brain and new verbal brain would be, as it was estimated in our time.

However, many scientists see the old thinking mode as dominated by a type of *correlative thinking*. The relative difference between correlative thinking and logical analysis means that the *ambiguity, vagueness, and incoherence associable with images and metaphors, which are specific for correlative*, are carried over into the more formal elements of thought that describe the logic. The correlative thinking involves the association of significances into clustered images, which define the context. **Correlative is a contextual "intelligence."**

One of the important devices for resolving the ancient correlation was the contrast of *yin* (shady side) and *yang (sunny side)*. ***Correlative descriptions are prescriptions.***

Since the old emotional thinking began losing ground against logical thinking, the **decrease in emotional began to elevate the power of concentration** that increases the *content-driven analysis* as a new intelligent behavior (as expressed in the today IQ scores).

There is a saying: you cannot be rational if you are too emotional, but you cannot be rational if you are not emotional. Your emotional brain represents your instincts, impulses, and intuition. Rationality represents some 20% of human decision making. The emotions still drive 80% of our choices.

Current statistics indicate the 25-30% of people use visual as the main mode of activity, 25-30% use verbal as the main mode of activity, while the remaining 40-50% use both modes. This information discloses that visual thinking is today, only 50% of total thinking. And the remaining 50% of the verbal activity of our brains takes place at a mental speed that is 20,000 times diminished compared to visual speed.

Overall, **at least half of our brain activity occurs at dramatically reduced speed** (approximatively 20,000 times slower). **This is our new verbal brain**.

A team of researchers led by Dr. Sebastian Goldt (University of Stuttgart, Germany) has discovered that "**the slower the neuron learns, it increases the neuron efficiency**." The study targeted a recent adaptation of the neurons in the human brain to use less energy for their functioning. **Slower learning diminishes energy consumption**.

I would say, this adaptation came in response to changes in the human diet (that seems to provide less energy to the brain), a reduction in physical activity, and a significant slowdown in the processing speed of the verbal neural networks.

However, some recent studies suggest that the hunter-gatherer's daily energy requirements were likely no different than those for current Western populations. The adoption of agriculture brought with it an increasing workload for Neolithic foragers because the older foragers enjoyed spending each day only a moderate amount of time on subsistence work. Thus, old foragers consumed less energy for foraging than the farmers do for land cultivation and their domestic work.

The foragers were able to provide more energy to their larger brains, which brains employed almost in exclusivity their visual but high-consuming-energy circuits.

As the language evolved, the verbal but slower circuits interfered with and/or overlapped some of the visual circuits (statistically about 50% of them). The verbal mode consumes less energy. But the introduction of farming increased body energy consumption, providing less energy to the brain needs. At this moment, the energetic balance of the brain was affected and probably forced to operate a neural adaptation to lesser energy provided to the brain.

It might explain why the size of the skull decreased more than 25% in the following few thousand years: gradually, the brain workload gradually diminished by 10%-20%-30%-50% because of a corresponding reduction in the neural working speed. Slower speed implies less mental work.

The verbal mode needs for neural circuits probably resumed to invade and take over on many existing visual circuits that gradually turned relatively less active, instead of building up their own verbal circuits.

As the process above continued, the development of the conjugated relationship in corresponding pairs continued, too. It has contributed to a significant increase in the overall activity of the left hemisphere, consolidating its dominant role.

As the left hemisphere activity increased, human language evolved, logic thinking was introduced, mathematics was rationally configured, imaginary-thinking flourished, thinking concentration increased, emotional thinking rapidly decreased.

The slower speed of processing of the perceptual information made humans see more details of reality, increased their power to discover. The perception of time gradually changed, too.

I assumed here that several neuronal mutations have occurred in the last 42,000 years. But the research in genetics seems to do not find any. This aspect may imply that **all forcing factors**, which intervened in this time interval, **have caused only a genetic new expression**.

However, as I assumed, ***everything that happened was driven by repeated stages of conjugation between the brain hemispheres***.

It is like having a house with only two rooms, and in each room, there are certain things. At some point, one started to move things from a crowded room into another but less crowded room. When this action ends, we still have the same house with two rooms, with overall absolutely same things, but the things in each room show a changed arrangement.

In sum, for an outside observer, absolutely nothing happened in this house because no new things were added or removed from it.

It probably is the case of our brain. **We still have the same Paleolithic brain** (because no significant mutations occurred), **while the neural circuits seem differently interconnected, favoring the connection between closer circuits**. Some brain closer sections turn very active, while the distantly placed sections lose much of the previous interconnectivity.

Then, what happened in this house? An outside factor increased the plasticity (or movability) of almost all things placed inside the house, and this plasticity/movability allowed the staff to be easily moved around.

Inside this house, the staff is absolutely the same, but because of the changes in arrangement, the things have now a new functionality resulting from a new type of interconnectedness.

Is that all? Not really. The changes inside the house have a big influence on the individual living here. In what sense?

The inhabitant becomes inspired to change the perspective on the practice performed outside because it becomes possible to plan, interconnect, and re-arrange in novel ways such as work, to predict and to obtain significant new functional benefits. Everything is now inspired by an "*imaginary-thinking*."

Even then, the power of imagination does not help know how the outside world is made. But, he/she could invent and construct an outside world that works only for itself.

In the end, one may have to confront the question: what would imply for our brain, if humanity would manage to (somehow) transform the verbal language into an exclusively visual (optic) language?

The answer may indicate that our brain would have to use high-energy consumption circuits, which would increase the overall brain consumption over its peak in the Late Paleolithic because now we add the complexity of language. The brain volume would significantly increase. We would need a much bigger skull (for a larger brain) that the largest skull that existed during the Paleolithic. And the neurons would have to re-adapt to higher speed processing. We would solve the problems much faster than we do now. All thing considered, such a future possibility of a visual (optic) language cannot be accommodated by our anatomy, except maybe in a very long-time interval of thousands of years, while it may imply the use of conjugates, which may not exist or not be available.

For the moment, we are constrained to rely on adding an exterior computing and problem-solving capacity, that is currently envisaged by an Artificial Intelligence. This situation is generated by the fact than we are exponentially confronted with an even-rising rate of artificial data produced by our civilization, which cannot be resolved by mental computation and investigation.

Here, there is no doubt that we tend to add an artificially created capacity to our little changed but Late Paleolithic brain. It does not matter how advanced this capacity would be because it would remain outside of our brain and way outside our biological thinking capabilities.

Even the mental-interference of a proposed interface-approach, being currently developed by the Artificial Intelligence program, cannot supplement the existing neural circuits in a truly biological way. Artificial Intelligence is not yet based on conjugated biological possibilities, which are naturally drawn to maneuver and accommodate our brain functioning.

In my opinion, for a very long time to come, the currently proposed Artificial Intelligence would remain separated from our brain biology and alien to ourselves.

I think that **current situation may have to prioritize the research** in the direction to investigate what other natural conjugates would be available for our brain system and mechanism, how could be stimulated their occurrence, and investigate what effects may generate such novel conjugates?

It may be the case that, in this manner, a lot of new but natural avenues can be unveiled, which may solve our problems outside of, while competing with the current Artificial Intelligence approach. Ultimately, they may help improve our sustainability, potentially resolving many severe environmental aspects.

> It must be a new while multidisciplinary field of competence. As Helena Knyazeva says in her article "*The Complex Nonlinear Thinking*" (2004): "All systems of human knowledge are based on certain principles. **The principles of complex nonlinear thinking bear in themselves the imprint of nature of principles as such**. Thus, the **principles of thinking should be holistic, complex and human oriented**. Theory

> of self-organization of complex systems plays a peculiar role in the indispensable reform of thinking".

Undoubtedly, the aforementioned hemispheric switching represents an act of self-organization incited by a forcing factor and/or a particular stimulus. It has produced a switching in brain behavior and functioning.

ATMOSPHERIC C14 ISOTOPE SHAPED MODERN BRAIN

Recent research (2012–2013) has determined that the era of testing of atomic bombs in the atmosphere (1955–1963) inflicted neuronal changes in the human brain for the generations who lived in that epoch.

The research has documented that, after the age of two years old, new neurons are formed in the human brain only in a section of hippocampus named *dentate gyrus*. Later research confirmed that this is the only area of the brain that experience new neurons after the mentioned age, and hence, it is prone to neurogenesis for some time during the adult human life.

Also, the research has determined that neurogenesis has a high impact on mutagenic substance and supports the production of mutated cells.

The researchers specifically found that the generations contemporary to the mentioned epoch of atomic testing in the atmosphere have been exposed to a higher concentration of atmospheric C14 than the subsequent generations.

According to literature, nuclear weapons testing in the 1950s and 1960s produced a level of bomb carbon content in the atmosphere that was 100 percent above normal levels when it was tested in 1963 and 1965. A normal abundance of C14 is one in a trillion.

During the atomic experiments, this level rose to two parts in a trillion (that is 100 percent increase). Such high level reached a peak in 1963 in the northern hemisphere, while the peak was reached in the southern hemisphere only two years later (1965) because of a slow atmospheric dispersion process.

Even after nuclear testing was banned, the bomb effect remained, and in the 1990s the level was only 20 percent higher than normal. In short, after ten years of bomb testing in the atmosphere, the C14 concentration diminished in thirty years to 20 percent from its peak. It means it took thirty years to diminish the effect by 80 percent.

We can consider that **the generation born between 1955 and 1985 have been affected and shows a higher level of C14 isotope in the DNA of their neurons**. As documented, those affected generations have a higher level of brain plasticity, and in general, that would imply a higher cognition capability when compared to the rest of population born before 1955 and after 1985.

The above is explained by a testing of the brains of the individuals who were exposed at birth, while documenting a generational differentiation in the functions of the dentate gyrus. Namely, the higher concentration of atmospheric C14 has stimulated the synaptic plasticity in the *dentate gyrus* for a limited time, meaning that **those affected *individuals become more readily adapt to the connections implied by new stimuli (environmental and otherwise stimuli) for a period at least equal to their exposure***, while the process could continue until the effect turns significantly diminished.

The researchers from Karolinska Institute (Stockholm, Sweden) and the German Center for Neurogenerative Diseases (Dresden, Germany) measured the concentration of C14 in the DNA of those particular neurons from the *dentate gyrus*.

It was found that ***the amount of C14 in the atmosphere is directly correlated to the amount of C14 existing in the Dentate Gyrus at birth***.

The research concluded that the vast majority of neurons in the human dentate gyrus (cca.70 percent) is subject to change compared with only 10 percent in mice and other mammals. Also, humans indicate a less pronounced age-dependent decline in hippocampal neurogenesis in contrast to mice.

But ***the most important conclusion was about the beneficial role of C14 in the dentate gyrus of humans that was not observed in other mammals***.

In my opinion, the mentioned generations born between 1955 and 1985 have a significant contribution, by an alleged higher power of cognition, to the exponential increases in the development of technology that manifested after 1980.

The research has established the link between higher C14 concentration on the neural DNA (in *dentate gyrus*) and higher cognition ability. It was specifically found that *this higher ability resolute in making better distinctions between the details of a perceived event*. I would say that ***such an ability would promote discoveries***.

When this enhanced ability helps separate some details from others, in fact, it *stimulates the analysis and favors memory storage of the detailed analysis*.

The researchers were surprised to learn that this ***new analyzing ability seems to stay longer in humans (at least twice higher) than in, for example, mice***. When the researchers experimented a similar C 14 high concentration on mice brains, they discovered that the effect on mice withdraws abruptly and after a short active period. The difference in the *rate of declining was ten to four* when the process was compared between the mice and human brains. Hence, in humans it declines 2.5 time slower than in mice.

However, the research teams found that the **neurons in the dentate gyrus of humans are performing better against this type of radiation than, probably, most of the other primates' neurons**.

The mentioned research and its results open an entirely new domain of scientific debate: the historic role of C 14 atmospheric higher concentrations in changing the human brain, which seems unique to humans (or better enhanced in humans than in other primates).

It is known that hippocampus is an evolutionarily ancient part of the cortex that connects with the neocortical area and is critical for the acquisition and retrieval of episodic and contextual memories. It is assumed that sensory information reaching the hippocampus via dentate gyrus (DG) granule cells is encoded in the CA3 auto-association circuits. Plasticity (as learning) is encoded in *memory traces*, while plasticity in Schaffer Collateral (SC) is necessary for *efficient retrieval* of such traces. In DG, the sensory information encounters a *pattern separation* process that is followed by encoding memories in CA3.

Neocortical gyrification is regulable and can be rapidly increased or decreased in evolution. There are indications that archicortical folding and surface area are regulable evolutionarily.

There are no studies that could define a temporal evolution and regression of the said process for periods encompassing longer periods, like thousands or hundreds of years.

But available data suggest that the capacity for well-regulated abventricular migration and proliferation of NSPCs evolved concurrently in DG and neocortex, increasing the cortical surface efficiently.

Most current hypotheses indicate that abGCs (which are new principal cells) *is for a period hyperplastic and hyperexcitable*. Eventually, the abGCs are the sole encoding units in the adult DG.

Here, the role of C14 in the neural DNA in Dentate Gyrus seems to have two ways of action:

- it stimulates the plasticity that produces an increased capacity of adaptation, and indirectly *it* ***stimulates humans to try new approaches*** *to which they appear adapted before-hand*; indirectly and directly, such stimulus ***favors "an augmentation of approaches***" or simply "augmentation";
- secondly, by increasing the flexibility, it stimulates a process of increased "socialization"; ***"increased socialization" also leads to "an augmentation of approaches"***;

Neurogenesis in mammals varies widely across species, with differences apparent in the *rate of proliferation (of neurons), survival (of the neuronal cells)*, and *rate of maturation (of neuronal cells)*. A great fraction of human GCs is subject to turnover, implying replacement and neurogenesis.

In my opinion, the *higher concentration of atmospheric C14 isotopes in certain epochs during the last fifty thousand years, probably affected the brains of several animals and birds, too.* The scale on which these animals were affected by C14 isotope remained small, but eventually, it favored some socialization approaches, maybe first manifested as self-domestication, while later it **allowed gradual socialization with humans**.

All animals and birds we have domesticated showed a tendency toward augmentation, meaning that they possess the capacity to grow in large numbers.

The mentioned *"regulation" of GCs* is known to be significantly made by *environment* and experience. Thus here, a high concentration of C14 isotope in the atmosphere will certainly influence GCs features, like their proliferation, survival, maturation, and the rate of mutations.

The behavioral attributes gained by humans during such epochs of high concentration of C14 isotope in the atmosphere may have turned into mutations. The ***cumulative process***

of adding new mutations has evolved in the direction of stimulating "an augmentation of approaches."

Adult neurogenesis increases the plasticity through multiple processes, like dendritic extension and retraction, synapses formation, and elimination.

Recognition of familiar objects (***pattern recognition***) and ***pattern separation*** are critical for effective memory, but a particular increase in such functions (like the growth of new axons, dendrites, and synapses) in the DG would almost certainly help produce something new, like a language.

Hence, one can presuppose that ***changes in human DG (under C14 influence) have dramatically contributed*** (as a development that implied several stages) ***to the onset of language*** (where after the incipient stage occurred some 70,000 years ago, allegedly the most determinant stage intervened some 30,000 years ago; but several consequent stages followed, like another important one was 4,500 years ago, while the final one was 2,700 years ago).

It might be so because ***DG volume predicts the accuracy and response time of the mentioned processes***. A large enough DG (even being a temporal feature) could help learning and memory, and from these features could arise the novelty of a language.

The variation (temporal increase-decrease) in the DG volume might be a likely explanation.

The fact that human hippocampus remains structurally plastic through life might represent a temporal variability that cannot be determined exactly for the interval of the last fifty thousand years.

As it appears, ***the increase of atmospheric C14 might have increased temporarily*** (for a period of fewer than three to six thousand years) ***the size of the DG***, pushing (I call it augmentation) all its functions on the edge that initiated the language production.

As it seems, the testing of the DG on cadavers of people who were born in the era of the atmospheric atomic experiments found an increased size in those persons DGs. The researchers also found that *DG volume is age-dependent*, the *CA3 volume is age-independent*, while DG inhibition, in general, produces loss of plasticity and memory.

The neural detail of the dentate gyrus (DG) seems to confirm a somehow distinct performance in humans. All other developed features, which are today so much advanced in humans, when compared to performances known in other primates, must collectively relate to *special qualities arising in the dentate gyrus (DG) of humans*.

Could enhanced plasticity (memory), pattern-recognition, and pattern-separation decisively contribute to language acquisition?

Could the same plasticity and selection-analysis make it possible for humans to rapidly adapt to all environments of this planet?

Talking about such a paramount role of the dentate gyrus, we do have first to define the variation of high concentration of C14 in the planetary atmosphere extended back to some significant time intervals (like in the last fifty thousand years) for human evolution.

But here, I have first to acknowledge that C14 lifetime halves in approximately six thousand years. Even then, the geophysicists tried to define this variability to the extent of last fifty thousand years using a combined set of high-precision Th230, U238, Pa 235, C14, and fossil corals.

The studies found an ***approximated relationship between geomagnetic field intensity and global C14 production***. It was found that, in general, the geomagnetic field intensity started to drop from fifty thousand years ago and reached an all-time low around thirty-seven thousand years ago. This all-time low intensity corresponds with the highest concentration of atmospheric C14 in the last seventy-five thousand years.

The ***geophysicists determined that the atmospheric level of C14 during a geomagnetic excursion could increase between*** 270 per thousand to 600 per thousand that is ***27 percent to 60 percent***. Normal C14 atmospheric level is one in a trillion.

The above figures indicate that the increase in the atmospheric C14 isotope during the excursions varies from 30 percent to 60 percent that is comparable to a 100 percent increase occurred during the atomic bomb atmospheric experiments. By contrast, the duration of the excursions was a minimum of one hundred years and reached a maximum (for this period up to fifty thousand years ago) of two to three thousand years.

However, the minimum intensity of the geomagnetic field, being recorded thirty-seven thousand years ago, eventually contributed to a larger increase in the atmospheric C14 isotope than the increase recorded by the geomagnetic excursions. The mentioned peak recorded thirty-seven thousand years ago was not an abrupt event, while it evolved toward its peak in the interval placed between thirty-eight thousand and thirty-six thousand years ago. Hence, here is another high concentration of C14 isotope of three thousand years.

I have to remind to the reader that the bulk of Out-Of-Africa migration left the continent in the same epoch of the last fifty thousand years. It can be associated with the era of increased brain plasticity, higher cognition, and better abilities for memory storage.

The language showed tremendous ability increases in the same epoch from fifty to twenty-five thousand years ago. The same era overlaps significant abstract art achievements.

Practically, the highest C14 concentration, occurring thirty-seven years ago, could be associated with overall completion of the migratory process, because those migrants were at the time in all Eurasia, inclusive in the distant Siberia, but also, they have reached Australia and New Guinea. Another aspect, to be mentioned here, refers to **the most important set of genetic mutations assumed in modern humans, which set develops in the same era**.

As the geophysical studies disclose, the atmospheric concentration of C14 followed an ever-downward path since its peak thirty-seven years ago.

Each of these excursions was a window in itself for galactic radiation penetrating to the surface of our planet, where the gamma-rays were the most influential generators of secondary particles of C14 in the atmosphere. Each shower of C14 had a highly active period of several thousands years.

In another paper uploaded on academia.edu (Augmentation of Brain Rhythms) I indicated that in the last forty thousand years human brain was subject to a *cosmogenic sequence of "cognition waves,"* and the role of the waves was to increase the entropy of particular sections of the brain but especially in the frontal cortex.

Now, the research mentioned above indicates a specificity associated with this process of high atmospheric concentration of C14 and discloses a determinant role of DNA concentration of C14 in the *dentate gyrus*. **The scientists were able to determine now the exact biological nature of the role played by C14 in the neural DNA**.

I like to discuss in some detail the duration of the C14 neuronal effect. The halving, as mentioned, is cca—6,000 years (more exactly 5,700 years). Thus, if Laschamp geomagnetic excursion that started the mentioned cycle of excursions of the last 50,000 years occurred 41,000 years ago, its sizeable influence would be a quart of the mentioned 6,000 years halving time. Such strong influence would have the duration of some 1,500 years.

Now, the sequence of geomagnetic disturbances manifested 41,000 y.a. (years ago), 38,000-36,000 y.a., 35,000-30,000 y.a., 25,000 y.a., 13,500 y.a., 10,000 y.a., 7,000-4,500 y.a., and 2,700-2,500 y.a. The last geomagnetic excursion, named Etrussia, theoretically should have produce a neuronal influence that is still active in human brains but was quite strong until year 1200 AD. Was this the indirect cause of Renaissance?

As one can observe, from 41,000 years ago to some 28,000 years ago, the concentration of atmospheric C14 was 30 percent to maybe 100% higher than normal (than normal we have now). This indicates a continual presence of higher than normal C14 concentration for at least 10,000 years that would correspond to the same size of continual increases in brain plasticity.

Now, I have to specify another element of particular importance. It refers to the nature of the "flipping" of the geomagnetic axis of the field. The geophysical axis normally mimics the magnetic axis that is known as the geographic north-south axis. In most of the cases, the flipping evolves for an inclination of forty-five to fifty degrees from the mentioned vertical display. Rarely, the inclination reaches the geographic equator, and even fewer times it passes south of the equator. This makes the case that galactic radiation would flood the area developed during flipping that is over the continental mass placed mostly in the northern hemisphere. The same flipping undoubtedly occurs in the southern hemisphere, but there it overlaps mostly some oceanic surfaces.

In sum, the continental landmass of the northern hemisphere has been traditionally affected by most of all geomagnetic excursions.

The result of such excursions was an increase in C14 atmospheric concentration. However, in a relatively short course of time, the atmospheric circulation has contributed to dissipate the high C14 concentrations in all directions and over the entire planetary surface.

Even then, the immediate and the highest impact was within the localities of the northern hemisphere.

There is a very well-defined anthropic contribution to increases in the atmospheric CO2 that is a complex of gases that include the C12, C13, and C14 among other gases.

An analysis of the period from 1750 to present indicates a steady increase of C12 and C13, but depletion of C14. Extraction and burning of fossil fuel release C12 but reduces C13. On the other hand, fossil fuel is virtually free of C14. This is so because plants preferentially take up C12 over C13, and therefore plants have low C13 in their composition. Through burning, the fossil fuels emission of C13 is lower than the regular concentration of C13 in the atmosphere.

Thus, the Industrial Revolution from the last two hundred years practically had no impact on atmospheric C14 concentration.

Even then, we should have some remnant higher concentration of C14 from Etrussia geomagnetic excursion that occurred 2,800-2,500 years ago.

Here, there is an aspect that may deserve some attention: C14 is an isotope produced by the impact of the galactic gamma-rays when crossing our planetary atmosphere.

But most powerful gamma-rays that could reach our planet originate into supernovae. Maybe a small discussion on the generally-admitted mechanism of supernovae may help make a virtual assumption on the fundament on which is based our civilization that in my papers, I call "augmentation."

A supernova phenomenon occurs when changes take place in the core of a star. Such a star is typically considered to have a mass six to twenty times larger (or even larger than twenty times) than our sun mass.

In the last stage, such a supergiant's core reaches significantly higher temperatures. Because of its core components of heavy elements, these elements absorb rather than release energy.

In the process, the outward pressure drops, while the inner pressure continues to increase that represents a collapsing. At some point, the collapsing material reacts to a core becoming increasingly rigid and bounces back from it into space in the form of a massive shockwave. This shockwave represents the energy of fifty times more the energy released by our sun in ten billion years.

The burst (called Gamma-Ray-Burst or GRB) of the shockwave can last from 0.01 seconds to more than one thousand seconds or sixteen minutes.

The type II supernova phenomenon is much easy to be observed, and this phenomenon occurs every twenty-five to fifty years.

Let's see what else supernovae produce. It is about the process named supernovae nucleosynthesis that fusion lighter elements into heavier ones. Here, the shockwave produces combined processes of nucleosynthesis and hydrostatic-burning from which result in most of the isotopes of carbon, oxygen, and others in the series from neon to nickel.

As a result of the ejection of newly synthesized isotopes of chemical elements, their abundance increases within the surrounding space, but it would cross into distant galaxies, reaching as far as the Earth. All heavy elements and metals on Earth originate into such ejections named GRB.

Somehow, the nucleosynthesis produced by supernovae repeats the nucleosynthesis that occurred during the Big Bang.

However, some scientists assume that radiation embedded into GRB-s must contain the coding of the entire history of a supergiant star from its formation to its final implosion, disintegration, and accompanying nucleosynthesis. In many respects, such supernovae history would mimic the Big Bang history.

As most specialists have agreed, "synthesis" is the process that generates or creates a system. Here, the "analysis" indicates the behavior of that system. These two notions can be used interchangeably.

In short, the GRBs are the bearers of the encoded message of "creation," where "creation" refers to producing every element of the universe, but where each element belongs to a system. ***The analysis shows how that system behaves, and the behaviors are embedded in synthesis***.

When the BGR-s react with our planetary atmosphere, it takes place the disintegration into secondary particles, and C14 and Beryllium 10 are such byproducts. Especially C14 is of great interest to human neurobiology because it was discovered to have a significant role in shaping our modern brain by enhancing brain plasticity, learning, memory, and cognition.

Even more important seems to be the fact that this type of features that do not appear in other mammals are developing uniquely in the human brain.

Usually, most researchers consider that GRBs may have a negligible mutational impact because of their short duration. But here, a long-term result is provided by a significant presence of C14 in the atmosphere and on the ground for a duration defined by a halving lifetime of 5,700-6,000 years.

The research found for the duration of cca. ten years of atomic testing an increase in atmospheric C14 to double than normal decadal figures recorded before and after this period.

In the end, I like to assume the potential role played by C14 in inflicting a type of "mental augmentation" in brains that is directly responsible for building our "augmenting type of civilization."

The language was the initial augmentation, and it opened a chain of many other aspects of augmentation, which all together define our civilization as a whole.

The documented (mentioned at the beginning of this paper) effect of *mental separation* between the elements of the perceived information is presupposed to be embedded in the concept of synthesis that came along with that information. *Trying to understand the behavior of that system of information, we found what we call analysis.*

In supernovae, the process is nucleosynthesis.

The mental analysis is a sub-product or a description of *mental synthesis.*

Such mental initiated procedures of *mental analysis-synthesis changed radically human thinking,* and it is associated with "*increased plasticity*" and "*memory storage.*" One might admit that plasticity is the process that allows understanding the meanings of synthesis (the creation) and analysis (the behavior of the created product).

In my previous papers, which I uploaded on academia.edu, I referred to an initiation of a novel process defined as "*imaginary thinking.*"

As researched, this "imaginary thinking" sometimes seems not related to any memory resulted from the experience. Thus, it generates an output that cannot be connected to any previous experience.

In its relatively rapid evolution, "imaginary thinking" turned increasingly disconnected from any direct experience, while its mental space exponentially increased against other components evolving in the mental space, like verbal thinking, thinking-as-talking, logic thinking (that evolved from analysis-synthesis).

"Imaginary thinking" is an alternative result of analysis-synthesis because it brakes memories (by analyzing) and synthesizes them into new constructs, which are very distinct from original memories. The result is about an "*invented*" product. The similarity to nucleosynthesis is consequential.

"Imaginary thinking" introduced more and more "artificial products, "which are "invertions" and which were nowhere in the material reality. I called this process the "*imaginary buildup of civilization.*"

However, "imaginary thinking" reflects a process of continual "*augmentation of everything*."

Returning to supernovae, the expansion of the superstar is a type of "*augmentation*" that defines its life. The superstar approaches a significant speeding up in the last phase, while its life ends by entering a collapsing process that is an "*implosion*."

As the supernovae show, such processes (augmentation and implosion) are fundamental in the entire universe. Hence, they are "*universal*."

Not coincidentally NASA/ADS physicists, who study supernovae, decided to replicate the astrophysical process in collapsing water bubbles experiments (named cavitation).

The cavitation experiments on water bubbles (drops) allowed the researchers to ensure the relevant consequences of the collapsing (implosion) process. They were able to visualize a jet pair escaping the drop after the collapse. The shock waves emitted at the bubble collapse bounce back and forth. The analogies with various astrophysical processes were established.

Wikipedia describes "cavitation" as a "phenomenon in which rapid changes of pressure in a liquid lead to the formation of small vapor-filled cavities, in places where the pressure is relatively low."

The *collapse is caused by those low-pressure spaces* or low density that means that a "cavity" form there. After the star collapses toward the core, it occurs a rebounding of the core that expels the outer layers in a gigantic explosion.

Taking the position of an exterior observer, no changes in the physiognomy of the star can be noticed until the shock wave, originating from the bouncing from the core, reaches the star's surface (envelope).

The shock wave rises enormously the temperature and produces a *dramatic expansion* (called explosion) *at a speed that increases exponentially until it reaches close to the speed of light.*

In the phase of collapsing, the star sucks an immense amount of matter from its surroundings.

In my opinion, the researchers who studied the role of C14 in the dentate gyrus of the hippocampus have validated the assumption that ***C14 may repeat on a self-similar scale the story of supernovae***

Here, I have to remind the reader that our brain selects something of one piece of information out of one million of such perceptual pieces.

We do not know if this selection leads or not to a self-similar product because our brain filters any replication of the universal processes. This filtering *dramatically alters the original information because the scale on which we transfer the information (that is the scale of our world) behaves with distinct physical properties.*

I assume that C14 in higher concentration would tend to mimic in our brains the steps of a "stellar type of augmentation," but which is an "induced" process that is not "replicated" in other animals.

It could be said that a *certain sensibility or resistibility to the effect of C14 isotopes generates an effect that uniquely manifests in human DG* for not fully documented reasons. It might be the case of a maladaptation of the dentate gyrus, or it may be something else.

This "mentally induced augmentation" makes us unique in the animal world.

However, our "mentally induced augmentation" *is also differential*, meaning that it produces "unevenness of development" or a "differential development." The unevenness principally refers to the main actor that the human-animal that biologically is very little changed compared to the Late Paleolithic era. The augmentation started in that mentioned

Late Paleolithic reached its apogee in around 1950 when it appeared a stringent need to replace normal brainpower with artificial automata, which opened the computers and Artificial Intelligence era.

This new era implied that brain plasticity reached its upper limit, where any further advances would be guided by automata. This is the discrepant point that pushes the differentially to an extreme.

Such "unevenness" is not physically sustainable, and it might lead to a theoretical collapse.

In sum, the relatively small difference between the processing in human Dentate Gyrus, compared to processing in mice and probably in all other mamals, have proved to represent a good case of sensibility to initial conditions. Thus, this relatively small difference produced a huge differentiation in the end result. This result was the human language that developed together with planning, control, and socialization.

The experiments indicated that, by inserting vitamin D and omega-3 fatty acid in the brains of irradiated rats, the severity of irradiation is greatly reduced. The same remedies were present in the prehistoric diet of humans.

PREHISTORIC HIGH CONCENTRATION OF ATMOSPHERIC C14 ISOTOPE

Part 1

The Reviews of Geophysics, Volume 57, Issue 3, has published on May 29, 2019, an article by J. E. T. Channell and L. Vigliotti under the title "The Role of Geomagnetic Field Intensity in Late Quaternary Evolution of Humans and Large Mammals."

I will quote from its Plain Language Summary placed at the beginning of this paper.

"The strength of the Earth's magnetic field in the past, recorded by rocks and sediments, provide a proxy for past flux of ultraviolet radiation (UVR) to Earth's surface due to the role of the field in modulating stratigraphic ozone. **About 40,000 years ago, mammalian fossils in Australia and Eurasia record an important die-off of large mammals that included Neanderthals in Europe**. In the Americas and Europe, a large mammalian die-offs can be linked to minima in Earth's magnetic field strength, implying that **UVR flux variations to Earth's surface influenced mammalian evolution**. For the last 200,000 years, estimates of the **timing of branching episodes in the human evolutionary tree, from modern and fossil DNA and Y chromosomes, can be linked to minima in field strength, which implies a long-term role for UVR in human evolution**. New and fossil find, improved fossil dating, knowledge of the past strength of Earth's magnetic field, and refinements in the human evolution tree, are sharpening the focus on a possible link between UVR arriving at the Earth's surface, magnetic field strength, and events in mammalian evolution".

The paper insists on the fact that **during geomagnetic polarity reversal, it also occurs a very low field intensity, and these factors combined have been the probable cause of large extinctions on Earth's biosystems**.

I would not enter the details provided by these authors on the biological mechanism affected by the UVR penetration on the ground.

However, I found it significant that such geophysical/ biological discussions appear now more frequent in the literature.

The scope of my paper is the discuss in more detail the idea of role played by the high concentrations of C14 isotope on evolution of the human brain and the approximative timing of such occurrences.

* * *

NTD Resource Center made an informative note that I will quote here.

"Cosmic rays enter the earth's atmosphere in large numbers every day, and when one collides with an atom in the atmosphere, it can create a secondary cosmic ray in the form of an energetic neutron. When these energetic neutrons collide with a nitrogen-14 atom, it turns into a carbon-14 atom and a hydrogen atom. Since Nitrogen gas makes up about 78 percent of the Earth's air, by volume, a considerable amount of Carbon-14 is produced. Carbon-14 atoms combine with oxygen to form carbon dioxide, which plants absorb naturally and incorporate into plant fibers by photosynthesis. Animals and people take in carbon-14 by eating the plants. Maybe one in a trillion carbon atoms is carbon-14.

Both Carbon-12 and Carbon-13 are stable, but Carbon-14 decays by very weak beta decays to Nitrogen-14 with a half-life of approximately 5,730 years. After the organism dies, it stops taking in new carbon.

Examples of the type of material that radiocarbon can determine the ages of the wood, charcoal, marine, and fresh shells, bone and antler, peat, and organic-bearing sediments. It also accumulates in carbonate deposits such as calcite, dissolved carbon dioxide, carbonates in ocean, lake, and groundwater sources, in stalactites of the caves.

Now, the recent summarization of geophysical studies published online by Cambridge University Press on 18 July 2016, indicate that new measurements for the paleomagnetic field strength, which show that the period before 12,000 years ago is characterized by low dipole moments or low strength. High values are associated with the Lake Mungo excursion between 32,000 and 28,000 years ago. In opposition to this, a field maximum has occurred 10,000 years ago and again 3,500 years ago.

* * *

Recent studies (2014) on a significant increase in the atmospheric concentration of C14 isotope during atmospheric testing of the atomic bombs found an important effect on the mechanism of Dentate Gyrus in the hippocampus.

The atmospheric tests of the atomic bombs, during the nuclear explosion, generated gamma-rays, which reacted with atmospheric Nitrogen-14 producing Carbon-14 isotopes.

The analysis and the experimental data of the mentioned effect suggest that, during that atmospheric testing era, the capacity for neurogenesis almost doubled as a result of doubling in the atmospheric C14 isotope.

There is known that during the geomagnetic excursions, the planetary shielding against cosmogenic radiation also diminishes in the range of 30-60%, allowing an accumulation of atmospheric C14 isotope 30-60% higher than normal.

However, the exact evolution of the mechanism and effect of each such excursion is not very well known, but this knowledge could be significantly improved. The radiative variation of each excursion can be better disclosed by a more comprehensive analysis of the geophysical and geological data at distinct planetary locations while employing novel correlations within other sources for C-14 data (tree rings, stalactites, sediments, pollen, corals, etc.).

Maps representing the relative path of the excursion could be drawn more precisely. Variation of the field intensity, along the length of the excursion at various locations, can be suggested by graphs.

Because of the aspect of the influence of locality during the excursions, the interpretation of maps and graphs would allow the specialists to establish quantitative values of geomagnetic field intensity at distinct locations during a particular time intervals that are defined along with the duration of each excursion.

In the end, each excursion would be represented by a map that contains graphs of each locality. The graphs would indicate the exact concentration of C-14 at a location and how it varied during the excursion.

Ultimately, the maps would indicate the variation of atmospheric C14 isotope at each locality along with the entire excursion interval.

Undoubtedly, such a project implies a tremendous work that would extend over many years.

However, the result would show what exact pulses of increased neurogenesis have occurred in each region and during which period.

Such a study could collaborate well with archaeological information available for each locality or an extended area.

In fact, here would be produced many maps, and the specialists would interpret the possible superposition.

The final result would indicate how different populations were or were not subject to such a radiative effect. It is important to know such information because of the large population mobility that occurred over the Eurasian landscapes.

We have today several planetary zones where the inhabitants display prehistoric types of culture. We call them "primitives," but it is essential to know if such "primitivism" may represent only a "blind-spot" on the radiative map. If this would be the case, then it means that we all have been significantly more or less affected by the radiative component caused by the combination of the geomagnetic excursions and the cosmogenic factor.

It will change our ideas about human and biological evolution on this planet.

Probably, it would change the way we see and understand the universe as a whole and our relationship with it.

In the last 40,000 years, the geomagnetic excursions occurred quite frequently, while they affected the landmass of the northern hemisphere, especially in the areas placed north of the 40 degrees northern parallel.

However, this represents the area were live and evolved the bulk of the planetary population and where the civilization evolved more rapidly than in other regions of our planet.

Hence, the suggestion made in this paper may help the specialists better understand what exactly shaped the evolution of human groups at various localities, and eventually, why the people have migrated so widely?

* * *

I like to present a short record and some data analysis available for the past 50,000 years on the atmospheric accumulations of C14 isotope.

Most studies accept the fact that C14 isotope concentration started to rise around 50,000 years ago, and increased continually until 35,000 years ago. After that peak, the C14 concentration took a diminishing path until nowadays.

The graphs used by these studies indicate that during the mentioned peak, the produced value of C14 was about 600-800 per million that represents an increase of 60-80%, and which is comparable with the increase of C14 recorded during the atomic bomb atmospheric testing that was 80-100%.

The atmospheric concentration of C14 was about 300 per million about 11,000 years ago (or 30% higher than current values), and it was 500 to 600 per million in the era 17,000 to 29,000 years ago (or 50-60% higher than current values).

It means that from 42,000 years ago, it had increased almost to double (or 100%) until 35,000 years ago, and it continued to remain high the entire era until 11,000 years ago (during the Younger Dryer cold epoch) when it still was 20-30% higher than current values. The field intensity suddenly increased only 10,000 years ago and for a short time.

After that increase, the field intensity dropped again for almost 5,000 years or until 4,000 years ago, but then, a new increase developed and reached its peak 3,500 years ago. In the last 2,000 years, the field intensity was constant.

However, this image of a higher concentration of atmospheric C14 for a duration of almost 30,000-34,000 years must imply a significant re-analysis of the evolution of the modern brain. This epoch stays in contrast with the previous era that occurred between 70,000-50,000 years ago when the C-14 concentration was low or normal.

However, the C-14 concentration was high again between 80,000-70,000 years ago that was the era of the first art expression at the southern end of Africa.

I would say that **these 30,000-34,000 years represented exactly the epoch when the prehistoric human brain changed into that modern brain that we have today**.

Nevertheless, based on other estimative sources, after the Lake Mungo phase ended (32,000-28,000 years ago), the excursion path continued with Mono Lake (28-26,000 years ago), Gothenburg (13-12,000 years ago), Solovki (7,000-4,500 years ago), and Sterno-Etrussia (2,800-2,600 years ago).

I like to specify that the time definition of each excursion varies from one source to another, and it is based on local geological and other information that is distinct from one locality to another.

Dipole intensity (field intensity) reached a minimum 5,500 years ago, during Solovki excursion, but rapidly increased back, reaching a maximum of 3,000-3,500 years ago. It dropped 30-50% again during the Sterno-Etrussia event (2,800-2,600 years ago), which lasted only 200 years.

It is assumed that a much more recent excursion occurred in the interval 774-775 AD.

The geophysics literature reports an extraordinary C14 large increase (20%) from 5481 BC to 5471 BC that is an event that occurred just 400-500 years before the starting of the Solovki event, and it was associated with a solar minima phenomenon.

All the above analysis of past C14 high atmospheric concentrations allows one to assume that, **besides the bulk of 60-80% increase, there occurred several other increases, which indicate 20-50% above normal C14 levels**.

It could be assumed that a case defining higher neurogenesis and increased plasticity in prehistoric brains developed in such epochs, which represented a unique radiative contribution for the entire interval of the last 250,000 years of human evolution.

It may suggest that the same processes contributed to genetic changes, as in the initiation of new haplogroups.

It may stimulate the very migration, too. It is possible that a deep investigation in such processes would unveil a mechanism that promoted mass human migrations beyond the known cause of climate change.

Other studies propose that the magnitude and direction of the geomagnetic field modulate the activity of cryptochrome (CRY) by influencing photochemical radical pair intermediates within the protein.

It shows that exposure to a magnetic field (100 mT) is sufficient to potentiate the ability of light-activated cryptochrome to increase neuronal action potential firing.

It proves that the activity of cryptochrome is sensitive to an external magnetic field, like the geomagnetic field variation, that is capable of modifying the animal behavior and causing changes in the gene expression.

PREHISTORIC HIGH ATMOSPHERIC CONCENTRATION OF C14 ISOTOPE

Part 2

(On academia.edu the original title is A new modality to investigate human brain evolution in the last 40,000 years). I like to mention and quote here two articles that I found only later and which provide precise data about the C14 isotope concentration from 54,000 years ago to the period of Classic Antiquity (8-3 Centuries BCE).

The first paper like to mention here is *"Analysis of the atmospheric C14 record spanning the past 50,000 years derived from high-precision Th230, U234, U238, Pa 231, U235 and C14 dates on fossil corals"*, written by Tzu-Chien, Richard G. Fairbanks, Li Cao, and Richard A. Mortlock, published in Elsevier (Quaternary Science Review 26) on June 19, 2006.

The study presents several graphs that I like to discuss in this short addendum to my original paper. In my previous paper, I was forced to make several assumptions based on lesser information found at the time in the literature.

The first graph I like to discuss displays radiocarbon production, tree-rings, Kiritimati, Barbados, Araki corals, and Araki and Barbados corals.

Here the graph indicates an increase in *carbon production* starting 48,000 years ago that reaches a peak around 41,000 years ago with a 60%-80% increase over normal values. Then, it drops to 23-25% 36,000 years ago, just to increase again around 34,000 years ago to 40%. It stays between 10% (33,000 years ago) and 32-33% since 22,000 years ago when it reached a new high of 38% around 22,000 years ago.

It remains between 35% and 10% until 8,000 years ago, when during Solovki excursion (7,000-4,500 BCE), it records another 28%, and another one during Sterno-Etrussia excursion (around 2,700-2,400 BCE) at 15%. In the graph, the *tree-rings* follow exactly radiocarbon production.

What is discrepant here, there were the values given by the corals. The corals show peaks significantly higher: for the epoch 42,000-37,000 years ago, and the values were 60 to 70%. And until 22,000 years ago the values remained above 50%. From 22,000 to 18,000, the values stayed at 45%. Between16,000 to 10,000, they were 45-48%, and 6,000 years ago, they reach values of 30%.

The other graphs of this paper show similar results.

The authors comment that the high C14 concentration decays slowly, while the increase is rapid when another excursion strikes shortly after that.

In sum, this paper indicates that the highest peak (almost 70-80%) is reached during the Laschamp excursion (42,000 years ago) and stay high for the next 3,000 years.

The second paper I like to mention is *"Atmospheric C14/C12 changes during the last glacial period from Hulu Cave"* (China) written by Hai Cheng, R. Lawrence Edwards, John Southon, Katsumi Matsumoto and a long list of colleges.

They have been able to define a more precise age measurement on several stalagmites in Hulu Cave (China).

They concluded the following:

- 54,000 to 50,000 years ago, the value was **12%** increased;

- from 50,000 to 43,000 years ago, the value increased another 14%, reaching a total of **28%**;
- the Laschamp excursion presented a short-term (temporary) reversal, and the reversal effects made this excursion significantly stronger than the following geomagnetic events, which randomly continued to occur for the next 30,000 years;
- from 43,000 to 38,000 years ago the values reach at its peak over **70%**, and on average it was **60%**, due to the high strength of Laschamp excursion;
- from 38,000 to 25,000 the values were **60% to 40%**;
- 20,000 years ago, the average value was **50%**.

In this first interval of 23,000 years long, the values were increased 50-60%.

The last interval occurred from 10,000 to 2,500 years ago:

- 11,000 years ago, the value was 15%;
- the value increased again 6,000 years ago (30%), and remained high for more than 1,500-2,000 years;
- the last increase occurred 2,500 years ago.

However, all this information confirms my statement from the original paper that **for at least 30,000 years, our ancestors lived within a high atmospheric C14 concentration that was in the range of 50% to 30% higher than the normal recorded before 1950**.

There is known that the activation of a wide number of differentiated signaling pathways for a population of neurons requires a high entropy.

Such activation defines a pluripotency state, where cellular diversity dominates, and it represents an undifferentiated/diversified state.

At a single-cell level, the high network entropy is thought of as a measure of the signaling pathway, where high-entropy would maintain open most options of the cell, which are associated with diverse cell fates. Such an undifferentiated state defines the network "plasticity."

Energy flow is fundamental for hippocampal dynamics that include the neurogenesis processes developed in the Dentate Gyrus placed in the hippocampus. Cross-frequency coupling is a manifestation of a nonlinear energy transfer.

Here, the energy of C14 isotopes is transferred to the neural network, increasing the entropy of this system of cells locally. This transfer gives a bust to the neurogenesis process in Dentate Gyrus and affects the fate and options of the newly generated neurons.

The research found out that hippocampal rhythm underpins language processing. This could explain why the high concentration of C14 isotopes in that the atmosphere in the era mentioned above has stimulated the human brain plasticity, contributing to environmental adaptation by higher levels of "plasticity" and the language development as another form of "plastic" adaptation.

Language is an undifferentiated complex, where "plasticity/flexibility" of the neural network helps adapt to various options, like communication, planning, control, and socialization. Higher "plasticity" offers more "options" that are capable of dealing with higher "environmental uncertainties."

All these mentioned elements contributed to the success of Out-of-Africa Migration, where the language was the core-technology employed to investigate and adapt.

There is no coincidence that those humans who went out-of-Africa cca. 50,000 years ago, get a big help cca. 43,000 years ago, to adapt to new environments by the intermediation of an increased atmospheric concentration of C14 isotope. C14 isotopes increased the entropy locally.

Here, there is to say that without such a lucky occurrence, the entire migration may have failed to succeed because there initially was not sufficient support to adapt to new environments.

There is known that an increase in neuronal network entropy increases oxidative stress and free-radicals. **De-excitation of free-radicals produces biophotons**. Short-pulses of augmented entropy generates biophotons and imply short-phases of biophotonic entanglement. There is to say that **such entanglement was the motor of correlative thinking, discoveries, innovation, and creativity, which all contributed to general human adaptation**.

The emission of biophotons, as a result of the de-excitation process, increases neural network communication, helping to increase cognition and memory storage, which further on participated in developing the language.

As it appears, during the mentioned era of low intensity of the geomagnetic field, pulses of gamma rays penetrated to the ground of our planet, and consequently, they generated pulses of C14 isotopes.

CHANGES IN BRAIN ANATOMY HAVE INFLICTED THE DEVELOPMENT OF LANGUAGE.

It is postulated, by several researchers, that the emergence of our species type of language-ready-brain is caused by developmental changes of brain morphology and neural connectivity. These changes give to our braincase a more globular shape.

I. Tattersall (April 2, 2019) suggests within his "*Minimalist Program and the Origin of Language*" (published in Frontiers) that **language emerged in a short-term event that occurred abruptly and late**.

He suggests but also finds it debatable that "**language was externalized in an independent event that followed its initial appearance**."

The focus on my paper refers to the fact that no archaeologist found an exact globular braincase dated before 35,000 years ago. Thus, I try to demonstrate here that the anatomical changes and alterations in the brain connectivity were both the result of a particular stimulus developing 42,000 years ago by a high atmospheric concentration of C14 isotope. This stimulus has been present in the Earth's atmosphere from 42,000 to 10,000 years ago, which is for 30,000 years.

However, geophysicists are aware of another 10,000 years long geophysical/radiative event that occurred between 84,000 and 74,000 years ago, and which event seems responsible for the first "art revolution" in South Africa. There is possible that this mentioned event had a certain influence at the beginning of language.

As I. Tattersall suggested, this was a configurative phase in language evolution that may follow up some other previous geomagnetic events. The **full externalization gradually occurred in an era that begins 40,000 years ago**.

In the meantime, archaeological evidence points to the fact that our braincase reached current shape and expression mostly during Holocene, around 8,000 years ago.

As it appears, the globularization appeared (35,000 years ago) shortly after the geomagnetic, and cosmogenic/radiative events turned manifested while affecting our brain biology. Still, the process continued to modify brain architecture until at least 8,000 years ago.

Niego and A. Benitez-Burraco, in a study on *Williams Syndrome* (published on March 18, 2019, in Frontiers) indicate that "**human language resulted from changes in our biology, behavior, and culture**. They found that William syndrome (WS) is a condition with a "clear genetic basis which results in a distinctive behavioral and cognitive profile, including enhanced sociability. In this paper, we show evidence that the WS phenotype can be satisfactorily constructed as **a hyper-domesticated human phenotype**, plausibly resulting from the effect of the WS hemideletion on selected candidates for domestication and neural crest (NC) function."

Wilkins at al. (2004) showed that "domestication usually entails a constellation of distinctive traits (the domestication syndrome), which are mostly related to the brain/cognitive changes. Such changes seem to be important for the development of language and globularity.

It has been considered that **different external stimuli triggered the domestication syndrome**. The result refers to the hypofunction of the neural crest cells (NCCs) in the embryonic phase of development.

It is known that **most adaptive differences** between the group of plants and animals **are due to changes in gene regulation** and not protein evolution. Genome-wide patterns of gene expression can evolve by a broader mechanism, where the non-coding variation (RNA) has a significant role.

It can be assessed that directional selection on RNA altered the expression of energy-metabolism-genes during domestication or self-domestication.

Different environmental conditions were manifested as stresses, while the gene activation came in response to corresponding biological characteristics, which are types of gene-expression.

Antonio Benitez-Burraco, Constantina Theofanopoulos, and Cedric Boeckx in a paper published on June 28, Volume 37, Issue 2, pages 265-278 in Springer Link considered that "the genetic changes we have claimed led to globularization are intimately related to those affected NCCs, fueling (or being the main rationale for) the emergence of the (self)-domestication syndrome in our species and ultimately, of full-fledged languages."

"The very nature of **cultural learning appears to be key in capturing all this grammatical paraphernalia** (Hurford 2011; Kirby 2012; Smith and Kirby 2008; Thompson et al. 2016). But Thomas points out, a major problem facing any attempt to account for language structure through a cultural mechanism is that the process required by such a mechanism is only possible if we assume the existence of preconditions, which we may call the "cultural niche."

"**Sensitivity to communicative intent**, at the heart of our communicative/pragmatic competence, **characterize domesticated species in general**. A cursory look at the fossil record highlight changes in anatomically modern humans that closely resemble the ones attested in domesticated animals: reduced skeletal and cranial robusticity, changes in dentition, retention of juvenile characteristics, changes in temperament (compared to other living primates), reduction of sexual dimorphism, etc."

"All these features suggest that **domestication may have taken place as part of the emergence of AMHs**. Although, of course, **once in place, it may have intensified in our recent history**."

It is a well-established fact that **domesticated animals possess a distinctive, unusual suite of heritable traits** not seen in their wild progenitors."

"We made as well an extensive survey of the PubMed literature to know whether our candidate (genes) can be regarded as "neural crest genes." The whole list of them, currently encompassing 152 genes, shows that 60% of them play a role related to the neural crest."

Here I suggest that the **correlation of several elements that contributed to the development of language**, and where one of them was self-domestication, **have been events that occurred in the same period 35,000-10,000 years ago**. It further led to another process, that is that of domestication of animals and plants, that seems to have started with dogs some 15,000 years ago.

Such correlations can also be drawn between cognition, culture, and art. The art explosion that dates some 30,000 years ago, also overlaps the language explosion and corroborates very well with a manifestation of human self-domestication, followed by the domestication of animals and plants. Here there is the origin of culture, too

Benitez-Burraco and C. Boeckx published on June 16, 2015, Frontiers in Psychology, a paper named *"Possible Functional Links Among Brain-and Skull-Related Genes Selected in Modern Humans."*

The authors reveal: "The genes we focus on are found mutated in different cognitive disorders affecting modern populations, and their products are involved in the skull and brain morphology, and neural connectivity. It means that the **changes affecting most of these proteins resulted in a more globular brain and ultimately brought about modern cognition**, with its characteristic generativity and capacity to form and exploit cross-modular concepts, properties most **clearly manifested in language**".

In my opinion, it shows that changes in brain morphology and connectivity changed the shape of the skull to globular and helped develop the language skills.

Modern behavior refers to creating tools and art, but which are secondary to language itself that represented communication, planning, control, and socialization. Scientists call such a period of their emergence the "**human revolution**" that started approximatively 50,000 years ago.

However, taking this advent into more detail, most scientists agree to reduce this interval to the last 40,000 years, while the globular features started to emerge only 40,000-35,000 years ago clearly.

In an article published in Science Advances in January 2018, Dr. Simon Neubauer of Max Plank Institute for Evolutionary Anthropology, in Leipzig, Germany, indicates:

"Globularity itself likely didn't give us advantages, but the features, that contributed to the rounding, probably did: **the bulging of parietal areas and the bulging of the cerebellum**. The parietal lobe is an important hub in connecting brain regions and is involved in functions like orientation, attention, and sensorimotor transformations that underlie planning and visuospatial integration. Meanwhile, the cerebellum relates to motor-related functions, like balance, as well as integral functions like working memory, language, affective processing, and social cognition. It's likely the emergence of these skills prompted the human revolution."

"Parietal areas are involved in orientation, attention, perception of stimuli, imagery, self-awareness, numerical processing, long-term memory, and tool use."

However, the **thalamus is central to language and human cognition**, as it modulates frontoparietal activity. There is here a gene-set, where its members are involved in the development and connectivity of the thalamus in the brain and produce language-associations.

"It is also interesting to point out that, in **present-day humans, brain globularity emerges** developmentally **during a few months around the time of birth** that is a critical and vulnerable period."

"**The globularity of modern man's endocranial cast is directly correlated with both hemispheres of the cerebrum, and both hemispheres of cerebellum-which make up the entire cranial globe**-becoming more like a spherical volleyball and less like an oval-shaped rugby ball."

"In the first three months of life, the cerebellum grows at the highest rate among other brain parts. The **characteristics of globularity seem to evolve during a prenatal and early postnatal period**".

Jean-Jacques Hublin, from the same researcher team in Leipzig, said:

"The **gradual evolution of modern brain shape seems to parallel the gradual emergence of behavioral modernity** as seen from the archaeological record."

Importantly, these **shape changes evolved independently of brain size**.

There is known that the human brain entered a size-reduction process that begins to manifest around 20,000 years ago.

Cedric Boeckx and Antonio Benitez-Burraco in another study (*"The Shape of The Human Language-Ready Brain"*) published on April 4, 2014, in Frontiers in Psychology explain:

"The **anatomical shift that led to globularity also entailed significant changes at the subcortical level**. Here **we focus on the thalamus, which we argue is central to language and human cognition**, as it modulates frontoparietal activity."

"It enables us to keep clearly separate two entities: one, the language-ready brain, understood as the cluster of brain properties that sets the stage for language ontogeny and phylogeny, and the other, language, understood as the collection of properties that humans eventually acquire as a result of social interactions."

"Our conclusion is also in line with more recent studies casting doubt on a direct link between laterality and language as a whole.

Thus, **to the extent that laterality bears on the linguistic brain, we think that it is likely to be tied to the communicative function of language**, or the externalization component".

"A detailed examination of endocasts from fossil specimens has revealed that modern humans, in contrast to otherwise heavily encephalized Neanderthals, show species-specific neomorphic hypertrophy of the parietal volume, leading to a dorsal growth and ventral flexion (convolutions) and consequent globularity of the whole structure. Subsequent research has established that **globularity is the result of a unique developmental trajectory in modern humans**, taking place at the stage of growth where the brain is the primary determinant of skull shape".

"**Globularity is what underlies our species language-readiness**." "Globularity de-emphasizes the role of frontal lobes in giving rise to modern cognition. The literature on globularity indicates that **at the very least parietal volumes are equally important**. Only Homo sapiens show a generalized enlargement of the entire parietal surface. It is indeed reasonable to think that the morphological **changes in the parietal region are to be related to important neurofunctional consequences, completing the functions of the frontal lobes**".

"Globularity is not just a superficial property of braincases. It crucially entails modifications of neural connections. The developmental trajectory giving rise to globularity is critical to the formation of a network of neural connections capable of supporting the most distinctive mode of cognition that numerous scholars have associated with language. Put succinctly, **a globular brain gives rise to the language-ready brain**".

So, the **frontoparietal-thalamic network emerges in the context of globularity**. As it seems, the dorsal thalamus, specifically the pulvinar and the mediodorsal nucleus, played a very significant role.

Here, **it occurred a coordinated expansion of functionally and anatomically connected areas, and in my hypothesis, such a complex process resulted from a highly stimulated neurogenesis in the Dentate Gyrus**. Such a phenomenon occurred in the interval 40,000-10,000 years ago, and it overlapped the mentioned chain of geophysical events when the atmospheric concentration of C14 isotope was the highest on a long geological scale (approximately 30,000 years).

I like to mention the results from some craniometrical studies, which enhance several other features produced by the process of globularization of the skull.

Such studies indicate that **reduced facial projection results from spheroid shortening** and generates other phylogenetically and functionally significant differences in craniofacial shape.

In the following paragraphs, I will refer to the research and the observations made by Daniel E. Lieberman with the Department of Anthropology, Rutgers University, at New Brunswick campus, and also at the New Jersey campus.

Hence, reduced facial projection increases overall cranial globularity by decreasing the cranial length relative to endocranial volume. It could explain how modern humans have evolved rapidly from more archaic forms. One possibility is that **a shorter spheroid, by decreasing the oropharynx length, has been an adaptation for speech, contributing to the unique proportions of the vocal tract, in which the horizontal component is equal in length to its vertical component**. This is in contrast with a markedly longer, as it exists, in other primates. Such a configuration **generates an ability to produce acoustically distinct speech sounds**.

However, the same reduction in facial projection helped humans to develop more significant **facial gestures**, which were initially used in sign languages.

Beyond the notion of involvement of the thalamus and the basal ganglia (BG) in linguistic operations, more specific functions of deeper structures remain controversial. In such a view, **thalamus** does not engage in proper linguistic operations. It **seems to act as a central monitor of language-specific cortical activities but supported by basal ganglia in all these activities**.

Brain-imaging methods revealed that much more of our brain in language processing than previously thought. We now know that numerous regions in every major lobe (frontal, parietal, occipital, and temporal lobes), especially in Broca and Wernicke areas, but also in connection with the cerebellum, are involved in the ability to comprehend and produce language.

Several authors have recently concluded that there exists a limited degree of spatial integration between soft and hard tissues, indicating that globularity and the development of various brain regions have mostly been independent events.

A very important aspect refers to **how the visual brain has switched to the verbal brain**, and if here, the globularization would have played a distinctive role or not? Then, the researchers "hypothesized that if the **visual cortex was involved in language processing**, those brain areas should show the same sensitivity to linguistic information as classic language areas such as Broca's and Wernicke's areas."

The last question was formulated by Marina Bedny, an MIT postdoctoral associate in the Department of Brain and Cognitive Sciences. Her response paper appeared in the Proceedings of National Academy of Sciences on February 28, 2011.

Her research found that the **visual brain regions were sensitive to sentence structure and word meanings in the same way as classic language regions**. She found that "these **brain regions can go from vision to language, indicating that the intrinsic function of a brain area is constrained only loosely**."

Amir Amedi, a neurophysiologist at the Hebrew University of Jerusalem, shows that the **left occipital cortex is processing language**. He said: "I think it suggests, in principle, and if the changes are forced early in development and early in life, any brain area can change its skin and do any task or function."

Bedny notes that "it suggests that a part of the brain can participate in language processing without having evolved to do so." She also thinks that this case occurs "as a result of a natural redistribution of tasks during brain development." "As these **brain functions are getting parceled out**, **the visual cortex isn't getting its typical function, which is to do vision**."

In sum, her study indicates that "different parts of the visual cortex get divvied up for different functions during development.

A subset of (left-brain) visual areas appears to be involved in language, including the left primary visual cortex".

Another paper by Sarah Genon, Andrew Reid, Robert Langner, Katim Amunts, and Simon B. Eickhoff was published in Trends in Cognitive Science in January 2018, under the title "*How to Characterize the Function of a Brain Region.*"

The authors show that a large-sample data can disclose covariation between brain region features and ecological, behavioral phenotyping.

"Current conceptualizations of brain function as a Bayesian machine, in which brain areas are seen as connected and relatively specialized computational units, are in contrast with actual available knowledge about functional specialization."

"We propose that assessing the relative functional specialization of brain regions requires a critical change in viewpoint, wherein the *a priori* defined construct is the brain region and the unknown are the behavioral functions associated with it."

"The cerebral cortex is not a homogeneous entity but can be subdivided into regionally distinct modules (cortical areas or subcortical nuclei) based on functional and structural properties. The latter emphasizes that **no brain region is by itself sufficient to perform a particular cognitive, sensory, or motor function**. Rather, **all mental capacities rely on a dynamic interplay and exchange of information between different regions**".

"Accordingly, each area can perform a limited range of functions, but the concrete behavioral output depends on which other areas (based on efferent connectivity)."

The study refers to a type of **coactivity** (as information is sent and received) **of each region and between regions**.

These new researches change our previous concepts about how the new verbal functions took over the old visual functions. In the meantime, it suggests that a particular increasing in neurogenesis have taken place in the recent past, and it contributed to a significant increase in brain plasticity, as it becomes manifested in various brain regions.

As it appears, the process of **overlapping of verbal functions on visual functions becomes possible by a significant increase**

in plasticity, and it probably does not diminish or annihilate the function it overlaps, as it was previously thought.

In support of the **idea of coactivity,** I like to quote another paper of Antonio Benitez-Burraco and Elliot Murphy, published in Frontiers in Behavioral Neuroscience on August 22, 2019, under the title "*Why Brain Oscillations Are Improving Our Understanding of Language?*"

The authors revealed: "It is now evident that **language results from the coordinated activity of several widespread brain networks**, encompassing different areas of both hemispheres. Oscillations enable the construction of coherently organized neural assemblies through establishing transitory temporal correlations. They reflect synchronized fluctuations in neuronal excitability and are grouped by frequency".

"For instance, they are primitive components of brain function and appear to be both domain-general (i.e., individual oscillations intervene in different cognitive and perceptual functions) and domain-specific (i.e., there exists a specific pattern of coupling between oscillations related to, and explaining each cognitive function)."

"**Thousands of biological factors interact to regulate language development and processing**. Also, brain oscillations are highly heritable traits, including oscillations related to language."

"Several linking hypotheses, connecting particular genes and oscillatory behavior implicated in language processing, can be posited, suggesting that **much of the oscillome is likely genetically-directed**; the set of genes implicated here is termed the oscillogenome. Importantly, these candidate genes map on to specific aspects of brain function, particularly on to neurotransmitter function, and though dopaminergic, GABAergic and glutamatergic synapses".

"A system biology approach to language is preferable since it allows us to understand how **language emerges from**

the complex interactions among thousands of biological factors, most notably oscillations. It is now clear that because language evolved mostly as a result of specific changes in the developmental path of the hominin brain in response to changes in the environment in which our ancestors lived, **we need to consider developmental, evolutionary aspects on a par**."

"The oscillations can explicitly be linked, in some way, to all major topics in the study of the computational nature of language. Specifically, oscillations might be a better (or perhaps, the optimum) candidate for properly defining the morphospace of the adaptive landscape of language growth in the species".

All oscillations do is to put together neurons and groups of neurons by a coactivation mode.

* * *

However, when one watches a map of the distribution of the language-related areas, one observes that all these areas surround the core of the brain where the thalamus is located. They seem interconnected with the thalamus and hippocampus.

Nevertheless, there, some philosophes debate the issue if we master the world by language, or it is the case that we only follow the language as it speaks to us.

Here, there is a debate about if the language is a representing, or as a propositional, or as an object. The answer to this debate is biased when it indicates that language would be an anatomical-technology, or it may not behave in such a technical manner.

For example, the Buddhists do not seek an experience that would transcend language, but they seek to free the language of its artificial boundaries. Those boundaries appear in our metaphysical way of thinking that sees the language as a tool for our use.

* * *

Kate M. Lesciotto and Joan T. Richtsmeier in their article *Craniofacial skeletal response to encephalization*, published in American Journal of Physical Anthropology, Volume 168, Issue S67, on 24 January 2019 explains:

"Another potential explanation for the globular shape of the human cranial vault has been proposed as part of a "**wiring" hypothesis, wherein the shape of the human brain is attributed to the need to reduce the wiring length** (distance between the axons and neurons to form functional circuits), both within the telencephalon and between the cerebrum and diencephalon (e.g., Bruner, Martin-Loeches & Colom, 2010; Chklovskii & Stevens, 2000; Hofer, 1969; Lieberman, Ross, & Ravosa, 2000: Mitchison, 1991; Ross & Henneberg, 1995, Sporns, Chialvo, Kaiser, & Hilgetag, 2004; Van Essen, 1997). <u>This idea</u> can be traced to Santiago Ramon y Cajal and is not specific to the explanation of human brains, as it <u>has been applied across species</u>. (Rivera-Alba et al., 2011; Stevens, 2012)."

"Some scholars have hypothesized that **a more spherical cerebrum would optimize neural connectivity of axons and dendrites within the evolutionary expanded hominin neocortex** (Bruner, 2004; Lieberman, Ross, & Ravosa, 2000; Ross & Henneberg, 1995). The location of the cerebellum and brain stem would anatomically prevent the expansion of the neocortex in a posterior or inferior direction, making anterior expansion the path of least resistance. To produce <u>a more balanced spherical shape alongside the anterior neocortical expansion, the brain necessarily develops a "kink" or ventral brain flexion</u> (Lieberman, Ross, & Ravosa, 2000)."

"At the time when brain globularity fell within the range of variation of present-day humans, the full set of behavioral modernity had been accumulated around 50,000-40,000 years ago."

I like to hypothesis the idea that <u>there had been a connection between language development</u>, especially manifested as the language revolution and the changes in the brain's anatomic architecture.

Definitely, the language brought to the brain a lot more information to be processed, and such a stimulus produced a neural adaptation in the brain network of circuits. Eventually, **the adaptation came in the form of shorter circuits around a central core of processing, and the adaptation was facilitated by a stimulus applied to that neurogenesis produced by the Dentate Gyrus**, also placed in a central position in the brain. As aforementioned, the stimulus was an increase in the atmospheric concentration of C14 isotope that started 42,000 years ago and lasted for the next 30,000 years.

The result of geomagnetic/radiative stimulus was a significant increase in the brain plasticity that was able to generate a new architecture where the circuits turned shorter, and hence, they became more efficient in dealing with that information produced by the rapid language evolution.

In my opinion, the change in architecture should be on par with the evolution of language. When we combine these two features or processes, the obvious result is a brain globularization.

"Language revolution," assumed to occur somewhere between 35,000 and 25,000 years ago, had produced a need for more energy: the processing of such language-related information requires more energy consumption.

As the archaeological skeletal investigation attests, another process developed in the last 20,000 years ago: it indicates a gradual reduction in human brain size.

However, the globularization of the skull does not correlate with the brain-size, meaning that the globularization process does not depend on the size of the brain.

As it can be assumed, the globularization reduced the connecting distance between the neural circuits, and probably it led, in the first stage, to a diminishing of energetic consumption.

In the meantime, globularization manifested by expanding several regions of the brain, like the parietal lobes, but also some centrally placed regions. These expansions meant new gray matter and new circuits, which altogether increased the processing capabilities of the brain.

The highest energetic consumption seems to be produced by the activity of memory. Storing memories increased as the language and social relationships evolved.

As I mentioned before, the language came within its evolving complexity, where the attributes of planning, control, socialization came together with the utterance. Thus, this complexity evolved in all its components.

The progress of the complex of language continued, and the initial small "saving" turned insufficient for a continual increase in neural energy consumption.

Overall, that prehistoric progress was significantly small when we compare with the anguage attributes existing around 2,000 years ago.

One can assume that this small "saving," obtained by shortening the length of the circuits, was exhausted around 20,000 years ago.

And then, another process started: a diminishing in the size of the brain.

All humans did was to learn. Hence, **learning and storing daily experiences became huge consumers of the brain's energy**.

How could this new process of brain-size diminishing be explained?

Obviously, it must be connected to the consumption of energy. Here were two ways to approach: one referred to a possibility to make the neurons learn slower, saving a lot of processing energy; the other way was offered by verbal communication in itself because the visual perception (that is a process of opticoding by opticoders) is much faster (probably at least 10,000 times) that processing of verbal perception (lexicoders).

So, in the first place, the **increased information produced by the language was balanced by its slower processing** within the neural network. For a while, this evolution was on par.

The most notable alternative to increase the amount of energy provided to the brain is to improve the diet.

However, various research that compared the Paleolithic diet to the current diet concluded that both are almost equal in energy production. Thus, the **introduction of agriculture did not change in a significative manner this energetic balance**.

It remains to disclose the way the neurons have turned to low energy consumption when confronted with learning tasks and memory storage.

Memory storage procedure shows that when one has to store something new, then, one has to erase something old or less important. Thus, the energy needed to store something new can be recovered by erasing an old memory. Here the balance is preserved. Nevertheless, this process does not influence the brain-size.

Now we came to the last element of this discussion: the slow-energy consumption of the neuron itself. As it seems, such stringency occurred in time starting 20,000 years ago and accelerated ever since until 8,000-6,000 years ago.

<u>Learning is a transformation of neural networks at the expense of energy</u>.

The above aspect was the subject of an article (*Stochastic Thermodynamics of Learning*) and research done by Sebastian Goldt and Udo Seifert and published on November 28, 2016, in Xiv:1611.09428v1.

The authors suggest that the thermodynamic cost of learning bounds information acquired by the neural network. The **organisms** (bacterium in the initial case study) **exploit a given energy budget to adapt to the environment**. Stochastic thermodynamics study the interplay of information processing and dissipation in interacting, fluctuating systems far from equilibrium.

"Learning is about extracting models from sensory data. In living systems, it is implemented in neural networks where vast numbers of neurons communicate with each other via action potentials, the electric pulse used universally as the basic token of communication in neural systems. Action potentials are transmitted via synapses, and their strength determines whether an incoming signal will make the receiving neuron trigger an action potential of its own".

The authors indicated that learning efficiency is constrained by the total entropy production of a neural network. They noticed that **the slower a neuron learns, the less heat and entropy it produces, which increases its efficiency**.

This study, in my opinion, suggests that there is the possibility that the human brain constrained by a high entropy produced by the language was forced to adapt by changing the efficiency of its neurons.

The adaptation, as mentioned above, was favored by a period of 30,000 years (from 42,000 to 10,000 years ago) when the concentration of atmospheric C14 was 60% to 100% higher than the current one. As a result of this high concentration, the Dentate Gyrus, in particular, was stimulated to increase its processes of neurogenesis significantly.

During the said period, it occurred an increase in the need for more metabolic energy to the brain. Hence, the **new neurons, as products of neurogenesis, came more ready to adapt to the imposed energetic task**.

I suggest that these **new neurons suffered the needed alteration that made them slow-learners**. In this manner, the **new neurons consumed less energy for processing the complexity of language**.

However, the efficiency of new neurons has led to the cancelation of those other neurons, which were no more needed. Probably, in time, the new neurons but highly efficient ones came in smaller numbers, causing a volume reduction.

This **volume loss refers to gray matter that diminished the brain volume as a whole**. **Previously, the shortened circuits have produced a reduction in the volume of white matter**.

There were two distinct processes, where both contributed to the general shrinking of the brain volume.

Consequently, the skull adapted, shrinking in the same proportion.

I am certain that there are many details of this process, which must be exactly established.

The globularization continued to evolve in the sense that it gradually affected deeper brain circuits, making them more efficient, shortened their length, and in the process, some more white matter (representing those circuits) became unnecessary due to shortcuts, and the neurogenesis did not renew it.

In my opinion, increased neurogenesis (that would refer to quantity but also the quality of those new neurons) has an essential contribution to all processes mentioned in this paper, being that fundamental factor that allowed such significant changes in our brain architecture.

Once again, our modernity started 42,000 years ago, and evolved gradually to current status, being uniquely intermediated by an era of high concentration of C14 atmospheric isotope that directly influenced the brain metabolism and neurogenesis.

ABSORPTION OF C14 ISOTOPE STIMULATES THE NEUROGENESIS AND BIOPHOTON PRODUCTION

In research, C14 (carbon-14) is used in carbonate form for isotopic labeling of molecules. Such activities use a dose greater than 1GBq, which means that the dose is around 0.4% (one in 250) higher than that provided by the natural amount of C14 isotopes in the atmospheric concentration.

C-14 is a low energy beta emitter, and even large amounts of this isotope pose little external dose hazard to persons exposed. The beta radiation barely penetrates the outer protective dead layer of the skin of the body. The primary concern refers to the possibility of internal exposure. The critical organ for most C14-labelled compounds is the fat of the whole body. Beta radiation has a short-range in the air.

When C14 decays, it emits a beta particle and becomes an N-14 isotope (theoretically reactive) with a half-life of 7.5 seconds. All other isotopes, but not C14, have a half-life of fewer than 20 seconds.

C14 needs almost 6,000 (5,730) years (half-life) to decay into a gamma particle and N14. After this decay, the remaining C14 and N14 are preserved within a 1:1 ratio.

There is a tremendous dilution of naturally produced C14 in the various pools of carbon in nature. In the atmosphere itself, the C14 is combined with CO2 making a radioactive compound with an atmospheric half-life of 12-16 years in the northern hemisphere.

C14 present in the atmosphere is mixed, and it is also mixed in shallow seawaters and freshwaters. In soils, as organically absorbed, it becomes associated with carbonate soil minerals.

According to literature, the excess carbon 14 produced during nuclear weapons testing has already decreased due to the global carbon exchange cycle. By the 1990s, the carbon 14 level was only 20% higher than the theoretical 1950 level of reference.

Various research estimates that, even on a time scale of a few decades, the atmospheric radiocarbon activity may not have been constant.

"The atmospheric radiocarbon concentration fluctuated due to natural causes by 2% during the Little Ice Age from the 16th to 19th centuries". "The abundance of atmospheric radiocarbon (C14) from 1500 to 1800 AD was 2% above average level, with a maximum that occurred for the years AD 1500 and AD 1700, and a minimum for AD 1600 and AD 1800". "Between AD 1500 and 1850, two maxima, two minima, and other minor variations are known as de Vries effect have been observed. The C14 maxima correspond to the Maunder and Sporer minima of solar activity". "This effect modulates the galactic cosmic rays' activity in antiphase with solar activity." (I quoted from *Temporal fluctuation of atmospheric C14: causal factors and implications* by Paul E. Damon, Juan Carlos Lerman, and Austin Long with the University of Arizona, Tucson, published in Annual Reviews, Inc., 1978).

The global carbon cycle refers to the exchanges of carbon within and between four major reservoirs: the atmosphere, the oceans, land (lithosphere), and fossil fuels. Carbon may be transferred from one pool to another in seconds (by fixating atmospheric CO2 into sugar through photosynthesis) or over millennia (the accumulation of fossil carbon through deposition and diagenesis of organic matter).

It is estimated (Nydal-1968) that the residence time in the stratosphere before transfer to the troposphere is 2.0 years. The transfer between Earth's hemispheres takes one year. The C14 atmospheric residence before transferring to the biosphere and aquatics is four years. Here, the residence in oceans is seven years, and from the ocean surface to the deep, the transfer takes 24 years.

Lingenfelter and Ramaty (1970) calculated that the solar flare of February 23, 1956, may have increased the atmosphere's radiocarbon concentration by 7.5%.

Paul A. LaViolette estimated in 2011, based on observations in the Cariaco Basin out of Venezuela coast, that a giant cycle of solar flares occurred in phases around 12,973 years ago, 12,337, and 12,639 years

ago. It extended on a length of about 300 years. The radiocarbon concentration rose by more than 50%.

A much smaller event was the Carrington event of 1859, but which was still 20 times greater than the 1956 event.

Another estimate indicates that Global Warming would cause a decrease in atmospheric C14 concentration.

"Because the industrial CO2 injected into the atmosphere has no C14, the effect decreased the C14 atmospheric concentration. This effect corresponds to the Industrial Era. It is estimated that C14 concentration before the industrial era was 2% higher than in 1950."

Konstantinov and Kocharov (1967) estimated a 0.5 to 40% increase of radiocarbon due to the gamma-ray burst from the supernova Tycho Brahe, seen in AD 1572.

Most such measurements were made on tree rings. The oldest bristlecone pine chronologies extend to 8,400 years ago.

The exchange time between the mentioned three reservoirs is ten years.

While the absolute sum of carbon found in the active reservoirs is maintained in a near steady-state by slow geological processes, more rapid biochemical processes drive carbon redistribution among the active reservoirs.

Dr. Zalasiewicz explains: "As a result of the detonations of hundreds of such weapons (atomic) around the globe, there will be plenty of these isotopes still around far into the future." (He refers to inorganic matter).

The amount of fossil fuel combustion decreases the C14 because fossil fuels have lost, on long geological scales, all C14 because of radioactive decay. In contrast, such fuel emission adds large nonradioactive volumes (CO2).

It is considered that today C14 concentration is 2% enriched compared to the preindustrial era (from over 200 years ago). The current trend in emissions continues to decrease the C14 concentration that, around 2020, it will reach the level from the preindustrial era.

By 2040, the concentration of atmospheric C14 will diminish by one quart compared to the preindustrial era.

Here, I like to note that such a decrease, from double concentration (around 1963) to 25% less than the concentration assumed to exists in the preindustrial era, may have vast mental implications.

However, such a change in C14 atmospheric concentration is generated by increasing double carbon dioxide at the same time interval.

It is well-known that it exists a direct proportionality between these gases concentration in the atmosphere: one component, like the carbon dioxide, rises, the other elements, like nitrogen (N), Oxygen, and C14, would diminish correspondingly.

For example, studies of euthanasia on mice have revealed the influence of such euthanasic gases on mice subjects:

-CO2 use increased the locomotor activity;

-N2 decreased such locomotion;

-CO2 increased the delta, theta rhythm correlation on the EEG, and increased excitation;

-N2 reduced behavioral activity and central neurological depression while being less aversive.

Both gases act on the brain and nervous system.

Other studies were concentrated on the effects of pollution on our neurological system. I will note several found effects:

-traffic-related air pollution (with mainly the carbon dioxide) exposure was associated with adverse effects, like a reduction in attention, global IQ, memory, and higher prevalence of several mental diseases (ADHD, ASD);

-it was observed a cognitive decline and increased dementia behaviors.

Global atmospheric CO2 levels passed 400 parts per million in 2016, double compared to the preindustrial era.

As the studies point out, the CO2 concentration in the atmosphere tends to increase almost to 1,000 parts per million toward the end of the current century.

However, we currently have frequent cases where the indoor concentration of CO2 is daily higher than 1,000 and even higher than 2,000 parts per million, especially in the school environment and almost in all office spaces. Even the concentration can reach over 1,000 parts per million in our houses when they are tightly isolated from the outdoor environments for energetic preservation reasons.

In sum, there are significant changes in the gaseous concentration in our planetary atmosphere, which changes affect our biology. Most importantly, they affect our brain workings in a very substantial manner.

It is hard to imagine that our Late Paleolithic brain would accommodate such functional changes when the previous source of plasticity that is C14 high atmospheric concentration tends to diminish significantly.

One last aspect is global nitrogen pollution on the rise,

but this subject does not enter the domain of issues I like to discuss here.

* * *

Let me continue with the role being played by C14 in the evolution of the human brain.

The animal organisms assimilate C14 Glucose when feeding on vegetables. This process stimulates the oxygen species (ROS) and regulates nitric oxide production and inhibition (NO), which both have interrelated roles in neurogenesis.

Under certain circumstances, experimental animal data suggest that the exposure give anti-tumor ability, slow progression of arteriosclerosis, and ameliorate diabetic nephropathy.

However, during experiments, the C14 traces have been 1 to 5mCi, thus entering the low hazard category.

The average concentration of C14 isotopes at sea level is 2mCi per year. During the atomic bomb testing in the atmosphere, such C14 concentration rose to 3.6mCi (as 80% higher than usual) as it was investigated in 1963-1965, and it diminished to 2.4mCi in 1990 (20% higher than average).

Thus, the atmospheric data mentioned above could be correlated with the dose used in pharmaceutical and medical experiments. The length of the exposure would produce fundamental differences.

The C14 isotopes inside the human body, as part of C14 Glucose, would stimulate the production of oxygen species (ROS), which have specific reactions with nitric oxide, regulating its concentration in various tissues. NO is known to create a nitrite reduction pathway. Within the blood vessels, the Oxygen produces vasoconstriction that causes hyperbaric tension within the vessels. High tension is also found inside the cells, inflicting the level of permeability of the cellular membranes. It increases the amount of available Oxygen in blood vessels and cells and interacts with local nitric oxide, regulating its production.

Nitric oxide is involved in transporting and dispersing Glucose in the entire organism. Glucose also stimulates oxygen species that metabolize any unwanted excess of this substance. When nitric oxide (as an antioxidant) decreases in the brain and nervous system, local oxygen species increases. By contrast, the same vascular tension affects the vascular peripheries distinctly, making the nitric oxide increase locally (vasodilation) while stimulating muscle fibers and skeletal developments.

The ROS regulates cellular differentiation, proliferation, apoptosis, cell cycle, and migration.

It is known that, during physical activity, the ROS and NO interplay stimulates oxygen production.

The total amount of ROS generated by primary ionizing events (like C14 isotopes activity when inserted into C14Glucose) is propagated to the level of intracellular activation. Almost immediately after exposure to ionizing radiation, the detoxifying enzymes activate the cellular defense mechanism of the antioxidant system. There is an entire biological mechanism that converges to repair nucleic acid damage and DNA modifications induced by the ionizing radiation, like the one generated by the atmospheric C14 isotopes when inserted into C14 Glucose produced by plants.

Low-dose Ionizing Radiation (IR) at 15-30mGy induces in biological molecules Reactive Oxygen Species (ROS) because gamma radiation of cellular water rapidly generates ROS like hydroxyl radical (OH) and ionized water.

The literature analyzed experimental cases where the irradiation of A549 cells induced mitochondrial ROS production, increased mitochondrial membrane potential and promoted respiration and ATP production.

The experiments with doses lower than 30mGy represent 4mCi radiation exposure, which produces an absorption of 0.4mGy/hr. proved to have no statistical significance. On the other hand, it may increase gene expression and trigger higher than regular mutation rates.

Experiments on mice colonies were prolonged to several generations, and the dose used was 4.5mGy/day to 28.8mGy at 1.2mGy/h. These experiments recorded many biopositive effects, like an increase in litter size, more fertile than the control group, increased litter number, increased viability, and faster growth rates.

However, a chemical reaction in cells caused by oxidation produces free radicals, which are the primary source of biophotons. Such biophotons are made during the free radicals deexcitation process.

Brain plasticity refers to an undifferentiated neuronal state. When the entropy occasionally increases in the hippocampus's Dentate Gyrus, a neuronal generation will burst with higher than average plasticity, meaning more undifferentiated neurons. It is known that higher oxidative stress generates a higher entropy.

Scientific literature estimates that biophotons would play a significant role in neuronal transmission and communication. Hence, the ionizing radiation inserted in C14 Glucose generated oxidative stress that resulted in a high biophoton production. These steps contributed to anatomical and functional changes in the Homo brain, producing a new type of mental activity and a modern intellect. The ultimate result was a revolution in language.

* * *

It has been experimentally demonstrated that a high concentration of atmospheric C14 has contributed to increased neurogenesis and higher brain plasticity when combining the interpolations.

The current trend of diminishing atmospheric concentration of C14 seems to have implications for the future human generations, making them less adaptable to changes.

This is important because we enter an Artificial Intelligence implementation era, where we will need a significantly augmented capacity of adaptation.

* * *

Another issue to be paid attention to is the ingestion of inorganic minerals by organisms, including humans.

Dr. Rick Wagner, C.N., explains in his paper Ionic Minerals:

"In nature, our mineral sources have always been the water-based either form of foods we eat or the water we drink. When minerals come from water, they are inorganic. When one ingests sea salt for the mineral content, they are utilizing inorganic forms of the minerals ".

"The important thing is that it does not matter if you consume minerals from plant sources, water-based, or solid pills or even dirt. What matters is how much time it takes the body to break down the minerals into their atomic (ionic) state to be effectively utilized at the cellular level. The body does not need its minerals bonded to carbon (ions are not bounded to carbon) for beneficial absorption and utilization. A mineral is an ionic form that alleviates the need for stomach acid to perform the function of ionization".

There is known that C14 usually bounds with sodium, potassium, and magnesium.

* * *

Returning to the investigative-use of C14 tracers, there is known that the human organism has a safety system that, in less than 48 seconds, eliminates a source that posits biological danger.

Such a trace-use of C14 isotope, as a beta emitter, is

low-dose equivalent and has insufficient energy to cause ionization but has enough energy to produce local excitation.

A radiotracer is a radioactive element, like C14 isotope, that is added to a nonradioactive part.

Various pharmacological testing and experiments with isotopic C14-labeled products (mostly C14 labeled glucose) resulted in an electrical stimulation that doubles the rate of glucose consumption. They showed a trifold increase in the rate of CO2 production.

These experiments indicated that the required energy increased as a direct result of increased respiration and glycolysis. Here, a supplementary metabolism was present, generating Oxygen, which was necessary to satisfy the increase in oxygen consumption.

However, the changes in the absorption of C14-labeled products by various cerebral structures resulted from their activation and the isotopic incorporation rate. When one substrate was involved in the mechanochemical reaction, the resultant product had the same isotopic composition as the substrate.

It was found that C14 is stably into organic molecules and blood.

Of interest to me, there was an increase of C14-glucose in the hypothalamus that generated an acute activation at the beginning of the experiments that was followed by a diminishing activation as a specific adaptation occurred.

Beyond the medical interest in the use of C14 as investigative tracers, it must be considered some other aspects of C14 isotope absorption by biological systems.

I have to specify that cosmic radiation, which produces atmospheric C-14 isotopes, currently generates a level of 0.3mSv per year at sea level.

During the atomic bomb atmospheric experiments, the mentioned above level was almost doubled to 0.5mSv. The same doubling occurred during the geophysical events produced by the geomagnetic excursions.

A person flying at 10,000 meters above the ground for 4 hours is subject to 20microSv. By comparison, a tracer used in medical and pharmaceutical fields can cause exposure of 0.2 to 5.2microSv, which is four times lower than the mentioned flight exposure.

When C14 decays, it emits a beta particle. The beta particle generates free-radicals. The presence of free radicals or reactive oxygen species (ROS) corresponds to ultraweak photon emission (UPE). ROS production plays a crucial role in defense against infections, apoptosis, aging, and cellular communication. The presence of reactive nitrogen species (RNS) that is the second element (as N14) encountered during C14 decaying also busts UPE or biophoton production.

However, one should consider that since a high atmospheric concentration existed for almost 30,000 years, such concentration (20% to 80% higher than usual) was preserved in inorganic systems at the same level for 6,000 years, each time it was produced.

The last known high concentration of atmospheric C14 was produced by the Sterno-Etrussia geomagnetic excursion of the geomagnetic field 2,500 years ago. It would remain present at its initial concentration in inorganic forms for 3,500 years from now on. It may continue to be absorbed from inorganic into biological systems.

There is information that neural theta rhythm (central to language production) is correlated with biophoton emissions.

The research has found that the biophotons are sent along with the axonal and neuronal ramifications and throughout the nervous system. Such biophoton conduction occurs with greater emphasis on the white matter of the brain representing the communication channels placed between the gray matter structures that define the neurons.

Reactive oxygen and nitrogen species (ROS and RNS) are also cellular metabolism products and have beneficial effects. The balance between prooxidants and antioxidants is critical for the survival and function of (aerobic) organisms.

* * *

The following part of this paper is dedicated to what the literature calls the critical role of Oxygen in regulating cell behavior during neurogenesis.

It intends to bridge the natural occurrence of high atmospheric concentration of C14 in our recent prehistory and a significant stimulation that inflicted human neurogenesis.

Joseph Tafur, M.D., Eduard P.A. Van Wijk, Ph.D., Roeland Van Wijk, Ph. D., and Paul J. Mills, Ph.D. in their paper *"Biophoton Detection and Low-Intensity Light Therapy: A Potential Clinical Partnership,"* published in Photomedicine and Laser Surgery, Volume 28, Number 1, 2010, Pp. 23-30, made a review of current knowledge on this matter.

Some of their conclusions serve very well the purpose of my investigation.

"We know that Low-Intensity Light Therapy (LILT) has the potential to accelerate ATP production and mitigate oxidative stress, which derives from excessive production of reactive oxygen species (ROS) or a lack of antioxidant activity."

"Cellular sensitivity to red and infrared light is influenced by the cellular redox state. The cellular growth phase, which may also correspond to the cellular redox state, appears to be another determinant of this sensitivity. Proliferating cells are, in many cases, more sensitive.

In each case, this proliferating phase is associated with elevated ROS production".

Here, the authors indicate that the proliferation of cells, neurons included, depends on the oxygen species' effect or redox. These effects are oxygen-dependent and involve the generation of ROS.

Wulf Droge, in his paper *"Free Radicals in the Physiological Control of Cell Function,"* published online on January 1, 2002, indicates:

"At high concentration, free radicals and radical-derived, non-radical reactive species are hazardous for living organisms and damage all major cellular constituents. **At moderate concentration**, however, nitric oxide (NO), superoxide anion, and related reactive oxygen species (ROS) play an important role as regulatory mediators in the signaling process. Many of the ROS-mediated responses protect the cells against oxidative stress and reestablish redox homeostasis".

Here, it is essential to note that C14 isotope, when its absorption occurs in a biological system, tends to decay into nitric oxide (NO) that would be lethal, as most scientists have previously thought.

However, this old thought significantly changed in recent years.

"Higher organisms, however, have evolved the use of NO and ROS also as signaling molecules to other physiological functions. NO, and ROS are typically generated in these cases by tightly regulated enzymes such as NO synthase (NOS) and NAD(P)H oxidase isoforms".

"In a given signaling protein, oxidative attack induces either a loss of function, a gain of function, or a switch to a different function. In mitochondria, ROS are generated as undesirable side products of the oxidative energy metabolism. The process of aging may result, at least in part, from radical-mediated oxidative damage".

"Free-radicals are important in biology because of their advantageous effects. In was discovered the role of nitric oxide (NO) as a regulatory molecule in the control of smooth muscle relaxation and platelet adhesion inhibition. A large body of evidence shows that living organisms have not only adapted to an unfriendly coexistence with free-radicals but have developed mechanisms for the advantageous use of free-radicals. The delicate balance between the advantageous and detrimental effects of free-radicals is an important aspect of life".

I quoted this author because he indicated how biology adapted to the production and presence of free-radicals.

Further, one will see that biology developed precise systems and reactions capable of de-excite or annihilate the free-radicals immediately after their mission was accomplished advantageously. Such a process prevents the occurrence of damages.

In the meantime, both papers quoted above indicate the stimulant. At the same time, the vital role of the oxidative stress that goes from increasing the cell/neuron proliferation (like in neurogenesis) to increased activation and signaling to the production of free radicals that, among other things, it seems to be the sole mechanism in charge with biophoton emission.

In an article (*"Oxygen, a Key Factor Regulating Cell Behavior during Neurogenesis and Cerebral Diseases"*) written by Kuan Zhang, Lingling Zhu, and Ming Fan, and published in Frontiers in Molecular Neuroscience, April 4, 2011, the authors explain:

"Oxygen is a significant substrate for energy production and cell metabolism. It is interesting to note that normal oxygen levels in the tissues are always substantially lower than 156 mmHgO2 in the air we breathe (Panchison, 2009)."

"The brain is one of the heaviest oxygen consumers in the body, which amounts to 20% of total oxygen consumption (Masamoto and Tanishita, 2009). The development of various organs of embryos, including the central nervous system (CNS), occurs in low-oxygen concentration (Fisher and Bavister, 1993. Chen et al., 1999). Apart from this, oxygen levels in brain tissues are often altered during stroke (Liu et al., 2004), brain trauma (Valadka et al., 1998), and in hyperbaric oxygen (HBO) environment (Balenane, 1982)".

However, hyperbaric oxygen therapy (HBOT) refers to the medical use of Oxygen at a level higher than atmospheric pressure (Jun Mu and John H. Zhang).

"Embryonic neurogenesis begins at the early gestation period under very low-oxygen concentration (15.2mmHg; Zhou, 2004)".

"Taken together, all of above researches indicated the significant role of the vessels as one of the important components of the neurogenesis niche, and may also imply that the higher oxygen tension around the vessels in the subventricular zone (SVZ) and Dentate Gyrus (DG) would be significant for the maintenance of the characteristics of NSCs."

"We found that the PO2 (O2-oxygen) levels in ventricles are in a dynamic state and fluctuate in the range of 42 to 48 mmHg at a frequency of about 3 minutes. In the hippocampus, the PO2 level in CA1 and hilus are very stable and maintain about 2mmHg, while the PO2 (Oxygen) in Dentate Gyrus is dynamic and fluctuates in a range of 6-8mmHg (Zhang et al., 2010)".

"The PO2 levels in uninjured brain tissue have been measured about 25-30mmHg in the white matter of frontal lobe (Sarrafzadeh et al., 1998), 20-40mmHg in normal tissue (Hlatky et al., 2003)".

"Proliferation was promoted, and apoptosis was reduced when cells were grown in lower O2, yielding a greater number of precursors. The differentiation of precursor cells into neurons with specific neurotransmitter phenotypes was also significantly altered".

"In summary, the moderate low-oxygen (15.2-38mmHg) concentration was able to promote the proliferation of NSCs from various resources and enhance the differentiation of NSCs into the TH-positive neurons.

It was found that the metabolism of cells would consume

the Oxygen of the medium".

I like to quote Amandeep Goyal, Tyler Chonis, and Jeffrey S. Cooper with their paper *"Hyperbaric Cardiovascular Effects"* published by Stat Pearls Publishing, January 2019.

"HBOT initiates generalized vasoconstriction of healthy blood vessels. Exposure to Oxygen's pressure of at least 2 ATA is known to induce arteriolar vasoconstriction and increase systemic vascular resistance. The primary mechanism that leads to this vasoconstriction involves a reduction of nitric oxide production in the endothelium. The hyperoxic environment leads to increased oxidation of nitric oxide (NO) radicals produced by the endothelium, leading to a loss of the vasorelaxant effect. Additionally, some research has shown that HBOT leads to alterations in other vasodilator compounds, such as prostaglandins, contributing to the net vasoconstriction effect.

Hyperoxia stimulates the sympathetic nervous system to promote vasoconstriction. Humans have been shown to have an augmentation of sympathetic nervous system activity."

"Although vasoconstriction partially impedes blood flow, the hyperoxygenation of the plasma results in an overall gain in delivered Oxygen. Short-term hyperoxia causes increased cerebral vasoconstriction and further the reduction of blood flow. However, even with the reduction of cerebral blood flow, the cerebellum receives more Oxygen than it would otherwise".

"The brainstem neurons play important roles in the cardioinhibitory center and underlie the hyperbaric reflex bradycardia. HBOT leads to parasympathetic activity and increases vagal tone. It has been suggested that another factor leading to the slowed heart rate is a nitrogen-dependent beta-blockade of the heart."

"Reduction of blood flow secondary to vasoconstriction leads to corresponding edema reduction."

Now, I like to connect the information about C14-labeled products and naturally occurred atmospheric showers of C14 isotopes due to the effect caused by the diminishing of the geomagnetic field intensity.

The geological samples, coral, tree-rings, and several other sources provide a record of the concentration of C14. In the past 42,000 years, our planet was subject to a high-frequency of short-bursts of gamma radiation, followed by showers of radionuclides, which we're incorporating the mentioned C14 isotope.

The C14 on the ground was absorbed by plants' photosynthesis and passed to animals and humans. Such absorption/incorporation process is similar to making C14-labeled products used as tracers for medical investigations. In the meantime, the C14-labeled products proved to act as biological stimulants, increasing metabolism and oxygen consumption. Such "increased oxygen production" from C14-labeled products was similar in many respects to "hyperoxia" also produced in the medical field.

In short, the release of Oxygen by neuronal processes exposed to the absorption of C14 isotopes generated a hyperbaric oxygen condition (vascular constriction).

Paul M. Macey, Mary A. Woo, and Donald M. Harper, of University College of London, United Kingdom, published an article titled *"Hyperoxic Brain Effects Are Normalized by Addition of CO2"* in PLOS/Medicine on May 22, 2007.

The authors show that "despite the objective of improving the tissue oxygen delivery, hyperoxic ventilation could accentuate ischemic trouble and impair the outcome. But several cortical, limbic, and cerebellar brain areas regulate these autonomic processes".

"We found, using functional magnetic resonance imaging, that 2 minutes of hyperoxic ventilation (100% O2) following a room air baseline elicited pronounced responses in autonomic and hormonal areas, including the hypothalamus, insula, and hippocampus, throughout the challenge".

"The addition of 5% CO_2 to 95% O_2 abolished the responses in the hypothalamus and lingual gyrus, substantially reduced insular, hippocampal, thalamic, and cerebellar patterns in the first 48 seconds, and abolished signals in those sites after that. Only the dorsal midbrain responded to hypercapnia, but this would be abolished, too, by additional CO_2".

At the beginning of this paper, I indicated that the pharmacological research with C14-labeled Glucose experimentally produced a twofold reduction of Glucose and a trifold increase of CO_2.

Here, a neural mechanism was unveiled, which automatically, within 48 seconds (to two minutes), would naturally reduce the effects of hyperoxia (generated by C14-labeled products and/or hyperbaric therapy).

It should be assumed that, during our recent prehistory, the same neural mechanism was in charge of reducing the natural hyperoxia produced at the time in the brains of our ancestors by the absorbed C14 present in their food and derived from a high concentration of this isotope in the terrestrial atmosphere.

Such a safety procedure was built-in into our neural network. It reduced the mentioned encounter to a series of short-term pulses, and where such short-pulses prevent the development of free-radicals.

I indicated that the concentration of atmospheric C14 isotopes was varying from 20-30% to 70-80% higher than usual for a period of almost 30,000 years in our recent prehistory (from 42,000 years ago to 2,500 years ago).

* * *

However, now I would assume that the past-occurrence of a high concentration of C14 isotope was producing a cerebral pulsing-process with durations of less than 48 seconds.

Such pulse-process, developed over 30,000 years, was

the natural stimulus affecting increased neurogenesis that shaped our modern brain into producing the plasticity necessary to create that language with which was buildup our civilization.

Here, it should be noted that an increase in the oxidative-stress produces free-radicals; the deexcitation of free-radicals generates biophotons.

In the case of C14 isotope infliction, the neuronal absorption process also produces oxidative-stress. And this last process, where the oxidation is termed hyperoxic, produces vessel constriction that mounts hyperbaric pressure; here, the nitric oxide radicals (NO) appear when the oxidation is longer than 48 seconds. The problem is resolved by a vessel relaxation that follows due to a subsequent increase in CO2 production in the brain.

There is possible that the C14 isotopes could generate oxidative stress, while two distinct processes annihilate the stress:

-the deexcitation of radicals that produces biophotons;

-increased local production of CO2 that relaxes the constricted vessels, terminating the hyperbaric stress and its oxygen production.

Suppose the hypothesis of a single cause for oxidative-stress could be proved. In that case, it may result that the same C14 infliction, in a particular era in our prehistory, generated simultaneously, but developed in two distinct processes, increased neurogenesis, and separately, it increased the biophoton production.

Here, the biophoton production can represent the source of a proposed neural optic communication. It could also serve as a fundamental for imaginary thinking that is activated during entanglement.

* * *

I like to bring to this end several quotes from two recent papers.

One of these papers indicates strong evidence that *"Long Course Hyperbaric Oxygen Stimulates Neurogenesis and Attenuates Inflammation after Ischemic Stroke"* (a paper of a large team of researchers from National Cheung Kung University, Kaohsiung University, Southern Taiwan University, all of them in Taiwan, published by Hindawi Publishing Corporation, Volume 2013).

The second paper shows that *"Hyperbaric oxygen therapy promotes neurogenesis"* (written by a research team from Loma Linda University, US, and Chongqing Medical University, China, published online 2011, June 27).

In the first paper, the authors said: "Several studies provided experimental evidence concerning the neuroprotection benefits of hyperbaric oxygen (HBOT) therapy and the influence of this therapy on the migration of BMSCs, neurogenesis, gliosis, and inflammation."

The second paper suggests that hyperbaric oxygen therapy enhances neurogenesis. The authors demonstrate the influence of HBOT on cellular transcription factors, including hypoxia-inducible factors.

To conclude, I would suggest a potential pathway for a high concentration of atmospheric C14 impact on the human brain in prehistoric times.

Atmospheric C14 is assimilated by plants during photosynthesis, helping the sucrose formation that is half Glucose and half fructose.

Further on, in this case, the Glucose becomes animals ingest the C14-Glucose along with the plants they consume.

In the human case, high C14-Glucose must increase nitric oxide production because it reduces blood glucose by uptake to skeletal and muscle systems. Ultimately, after a pulse of increased production, it occurs a general reduction of nitric oxide in some of its components.

Lesser nitric oxide generates the vasoconstriction that increases the oxygen tension in the cells and increases the neuroblast proliferation (neurogenesis).

Vasoconstriction in the brain stimulates neurogenesis, and consequently, brain plasticity and cognition.

The decrease in some nitric oxide components produced an increased resistance to viruses and pathogens. Such novel immunity helped with the migration out of Africa and adaption to new ecological conditions.

IMAGINARY THINKING

Imaginary thinking has its roots in Magical Thinking. As Alex Lickerman, MD, (Happiness in this World), defines it in his paper, Magical Thinking (posted on November 12, 2009, in Psychology Today):

"Magical thinking is defined as believing that one event happens as a result of another without a plausible link of causation or even that an effect can occur without a cause at all. Perhaps, then, a more nuanced definition would believe in things more strongly than either evidence or experience justifies."

"We could conclude something is true for which there is only subjective evidence or experience (meaning, not objectively demonstrable to anyone else). For example, a vivid memory of a past life is a memory that you could question its validity but not the thought process that created the belief."

The thinking is a process that relays on memory storage, and memories are all past experiences. The use of language helped communicate and share individual and even collective memories. The language found a way to code such memories in vocal symbols.

Before the vocalization got coherence, for a while, it existed the sign-language that seems to have in it some incipient grammar rules. This implies that unstructured data became semi-structured, allowing further processes of increase structuring.

We can assume about magical thinking from the burial traditions of the Paleolithic people. Around thirty thousand years ago, we found the "magic expression" in the abstract symbolism inserted in the artistic expressions.

However, a civilization evolved by structuring the unstructured data. But semi-structured data, or half-processed data, is such data that does not fit into the relational model where structured data exists.

The reason we structure the data is that we need to place the structures in a relational model, where analysis can correlate between structures. The structured data could be experienced and turns stored in memory.

The semi-structured data appears like existing subliminally because it has the appearance of memory shards or deconstructed memories, where the shards are rebuilt into new mental constructs.

Semi-structured data comes from instances where data does not have rigidity and predefined schema. But also, it tries to make a flexible data cope with changes in relationships between data records that maintain the structure. Here, the impact of analysis-synthesis is clear.

Now, imaginary means a new type of grouping of data, where the rigidity of the experience record is broken down and rebuilt differently.

The experience is accomplished because the thinking flexibility allowed a higher capacity for adaptation. It incites one try new approaches, which would lead to new experiences because the thinking flexibility assures adaptation to some unknown approaches. This is like a before-hand guaranty into succeeding new approaches.

A certain thinking flexibility is a stimulus toward new approaches. But what are such new approaches? They must be different from whatever have been already experiences. They imply a type of imagination.

Even then, the imagination must be constructed from something that is already known because it has been experienced and is recorded in memory.

The memory contains all elements, which can be used in imagining. And, the memories must be analyzed, meaning breaking them into component pieces. The analysis deals with behaviors of a system, and memories are systems, too.

However, the memory pieces are now very distinct from the original storage of experience because those pieces express a type of self-similarity, meaning that the pieces look similar to the original, but their properties are different.

Synthesis is the process that reconstructs together the self-similar pieces, and indeed, the final construct would have distinct properties.

Humans, having increased thinking flexibility, needed gradual changes, and various adapting behaviors.

In the end, the imaginary thinking generates a product that represents a "novel" approach to the existing products; some may call it a "discovery" product or an "innovation" product.

The success of such innovations stimulated the development of "imaginary thinking" while ensuing an ever growth of this system of thinking and its "product."

One could say that such products have a "magic" in themselves, while they represent an "artificiality" compared to natural products. This "artificiality" originates into analysis-synthesis process, where the elements of the behavior are recomposed into new behavior. The new behavior is the sum of the decomposed parts, and this "sum" is never equal to the original but natural product that was analyzed.

Another article, that I wrote about, "The role of Atmospheric C14 isotope in shaping modern brain", showed that several research studies concluded that the "C14 concentration in genomic DNA of hippocampal neurons reflects the magnitude of neurogenesis manifestation."

In short, **higher C14 isotope concentration produces increased neurogenesis**; increased neurogenesis generates higher brain plasticity, higher adaptability, and higher cognition. All these new characteristics (behaviors) could turn into neural mutations, changing the brain functioning.

As Gerd Kempermann, Daniela Gast, Golo Kronenberg, Masahiro Yamaguchi and Fred H. Gage explained in their paper Development, published in 2003, an increase of "functional role of adult hippocampal neurogenesis is to enable an increase cooperation to "novelty" and "greater complexity" because such type of increased "neurogenesis leads to fast, stable and, lasting integration of new neurons."

As mentioned in my articles, the main functional role of the dentate gyrus of hippocampus refers to "*pattern separation*" from perceived inputs. This process of separation evolved into "analysis and synthesis," driving human thinking into logic approaches.

In the mentioned article, I indicated that a new cycle of geomagnetic events, occurring in the last forty-two thousand years is responsible for temporary increases in C14 concentration of neuronal DNA in the hippocampus.

Dr. Richard Klein, a paleoanthropologist at Stanford University, suggested that a genetic mutation occurred forty thousand years ago and caused an abrupt revolution in the way people thought and behave.

The mentioned date represents the ignition of a cycle of biological **changes in the human brain, which all but gradually *favored the development of "imaginary thinking."*** I also called this gradual development under the name "augmentation."

In another article uploaded on academia.edu under the title, "Augmentation of Brain Rhythms," I explained that the same mentioned cycle of geophysical events (named geomagnetic excursions) generated a *temporary increase in the brain rhythms* that favored increased cognition. For this reason, I called them the "waves of cognition."

Thus, the scope of this paper is to define the complexity that represents the "imaginary augmentation."

I will start with the capacity to "mentalize." Its asses self-reported ability to recognize and react to others' emotions and to interpret them, while having the task to identify what is expressed by visual perception.

The research has identified the meanings of "mentalizing" (measured by Empathy Quotient) by studying cases of "mentalizing deficits," like in subjects with the high score on Autism Spectrum Quotient, Schizophrenia, and Asperger syndrome. People affected by these diseases tend to do not believe in God as a person, but may accept a superior rule that makes things orderly, and that is similar to the thinking of the atheists.

There is the belief that "mentalizing" is a production of consciousness that is intimately related to the development of the ego.

On the other hand, "mentalizing" seems to have been evolved out of "magical thinking." The conclusion was that "mentalizing" tend to attribute purpose where there is none.

Let's see what daydreaming is. As psychologist Eric Klinger of the University of Minnesota explains, it defines the daydreaming as the time when "the mind floats in unfocused mental space." "An increase in imaginative daydreaming *replays variants of the millions of events stored in our brain. Allowing our mind to ramble, it enables us to access ideas not easily available to our conscious mind or to combine these insights in original ways.*"

Now, I like to quote Peter Reueli (Harvard Staff Writer, October 6, 2015) article, "How the brain builds new thoughts."

"One of the big mysteries of human cognition is how the brain takes ideas and puts them together in new ways to form new thoughts," said Steven Franklin and Professor Joshua Greene in their study from September 17, 2015, published in in the Proceedings of the National Academy of Science.

"How are such thoughts constructed? According to one theory, the brain does it by representing conceptual variables, answers to recurring questions of meaning. A new thought can be built by making "beating" the value of the "patient" variable. Franklin and Greene are the first to point to specific regions of the brain that encode such mental syntax."

"To identify the regions, they used functional magnetic resonance imagining (fMRI) to scan students' brains." "What we found is there are two regions in the left superior temporal lobe, one which is situated more toward the center of the head, that carries information about the agent, and one doing an action. An immediately adjacent region, located closer to the ear, carries information about the patient, or who the action was done to. The brain appears to reuse the same patterns across multiple sentences, implying that these patterns function like symbols."

"It suggests that these symbols are used over and over again to compose new thoughts. We find that the structure of the thoughts is mapped onto the structure of the brain systematically."

> "That ability to use a series of repeatable concepts to formulate new thoughts may be part of what makes human thought unique-and uniquely powerful."
>
> "We have this library of concepts that we can use to formulate an effectively infinite number of thoughts. Humans can read or hear a string of concepts and immediately put those concepts together to form some new idea."

Ultimately, higher cognition relies on the capacity of the individual to combine conceptual building blocks, which would design new thoughts.

Here, there is gradual progress where the individual plays around with images and concepts, which become fine-tuned, reconfigured. An exploratory play is needed to create a diversity of experiences or ideas. Only some of them would be a success, while the rest is discarded. In general, we are failing more often than we succeed.

There are plenty of studies and experiments that prove the importance of "imaginary thinking."

The things go from "pretending play" in childhood, to imagine about thinking positive, pretending to do an exercise, pretending to play an instrument or simply pretending eating, or pretending anything else.

Whatever we pretend, as it seems, it would probably occur in our mind, because our mind would make all possible associations also pretending that that thing happened.

We know that, when the context changes, the meaning of the content changes, too. Hence, when one thinks about the desired content, his/her mind will try to filter the perceptual input until there would find a mental context where that desired content may exist. So, the content would exist only in a selected context.

Time and again, the language were the tool that stimulated "imaginary thinking," and this type of thinking was able to create a type of imaginary reality, where all products are artificially built.

All inventions were first imagined. The man became accustomed to inventing "imaginary understanding" and accepting it as belief reality. In the last five thousand years, we begin to produce "factual reality." We start viewing logic as imaginary, and imagination as reality.

This is so because we analyze the behaviors of reality, and then, we synthesize pieces of behaviors into an "understanding" that is an "imaginary understanding" and which represents an "understanding of behaviors."

In all that, we use our own mind context, and the behaviors are implied by that context. Changing the context, our reality will behave differently.

The mentioned context is entirely dependent on our brain filtering or selection of perceptual input. We allegedly select one-piece of information out from a million pieces. Such an accumulation of many one-piece forms our context and shapes our understanding.

I like to mention that nootropic agents have been used for at least ten thousand years as documented by archaeological research. Many scientists see a quite significant role in the prehistoric and historic use of nootropic agents, which allegedly contributed to significant increases in the human imagination.

According to written information from Antiquity, these nootropic agents were not known to develop any dependency or negative effects.

A recent investigation of higher concentration of C14 in the neuronal DNA, as generated by atomic atmospheric testing in 1955–1963, found that the effect of C14 was, among other things, to protect the individual against addiction and other negative effects known to be currently produced by drugs.

It will be interesting to mention the era when, in the US, the drugs were not prohibited by law. During that era of the 1960s, the C14 atmospheric concentration was high and prevented to develop negative effects for those who do not abuse or use excessive-high drug doses.

As the C14 effect gradually withdraws after the banning of the atmospheric tests, the effect of the drugs significantly increased and led to their outlaw since 1969.

Here, one can assume that a high concentration of atmospheric C14 isotope dominated around 10,000 years ago, and again between 7,000 and 4,500 years ago (during Solovki event). One last geomagnetic event occurred between 2,700 and 2,500 years ago.

In these three periods of the last 10,000 years, a popular drug consumption (mostly in the form of mushrooms) will not lead to observable negative effects on users.

As the research has discovered, the using of meditation procedures also stimulates the imaginary thinking and does not imply any negative effects.

Since 1960, it developed a very rich literature about the "imaginary institution of society," hence, it would be inappropriate to initiate a discussion here.

I like to quote Prof. Stavros Moutsios' paper, "Imaginary Significations and Education as a Social Institution" (October 2013), "the functional/instrumental aspect of social institutions follows the operative schema of determinacy: something exists because it can be defined and determined. Imaginary significations can be identified but not determined, nor they can be reassembled or reproduced. They are in a constant state of transition from flux to structure." "Culture is embedded in the imaginary institution of society. ***Each society is instituted by***

creating its own world of imaginary significations and exists through them."

In the end, I like to specify that "augmentation" is a phenomenon that refers to biological systems, and it is unrelated to biological growth that arises through cell division. I would say that ***"augmentation" is a type of stimulated complexification that is encountered by a biological system***.

Our civilization, by all means, ***is an augmentation of the imaginary***, while it originated in the development of language that served as its sole vehicle of expression.

But this subject of phenomenon of augmentation will be separately discussed into another paper.

PHENOMENON OF AUGMENTATION

The phenomenon of augmentation is a natural outcome. It is not caused by evolution, while it appears as a result of a particular stimulation inserted by other universal phenomena of the Universe, which seek their own replication.

The process of augmentation affects a whole species or even a group of species. It is not related to biological growth. Like every other aspect of the natural world, augmentation proceeds with developing multiple stages of complexification due to the occurrence of multiple bifurcations.

I propose that "augmentation" varies in its development from one case to another, but its manifestation represents a reduced scale on which a star history is replicated. I said a star because the stars are the main elements of the known universe. Here, the particular case is a supernova, and its outcome is embedded in the ejection of gamma-rays.

How the supernova history would be embedded in the gamma-rays ejection is a problem that would be explained by quantum mechanics. As a very general idea, quantum physicists say that two particles, which have a common origin, would always be interconnected. In my opinion, this is a way to transmit, while helping to replicate the same story all across the universe. Incidentally, this same story was originated in the Big Bang, while the supernovae replicate it on a significantly smaller scale.

Again, in my own opinion, all different expressions took by the processes of "augmentation," are in fact fractal replication of supernovae's story. The supernovae ejected gamma-rays are the vehicles transporting such stories.

* * *

Complexification is described as an *incremental elaboration of solutions by adding new structures*. In a plain language, every system of the universe encounters problems, and complexification is the needed answer, providing solutions.

In nature, evolve competing for elements and structures, which give rise to systems. Natural competition causes local problems, and naturally, some fitness solutions are naturally proposed for each system. The solutions are provided and implemented by the complexification processes. But complexification does not stop because the solution fits only temporarily, and a bifurcation follows up that is another search for a better solution.

However, the science of complexity indicates that a system during complexification tends to bifurcate, and each new search-path bifurcates anew. Each path is a new search path for solutions, but too many bifurcations increase the entropy (disorder) of the system until it loses the coherence (orderly behaviors) and collapses.

The elements resulting from the collapse restart the complexification process. This process repeats itself on and on. The collapse is the dissipation of complexity.

Thus, the ***complexification only tends to bring a solution with a certain temporality***. By contrast, ***augmentation tends to replicate the original mechanism that created and creates everything***.

At the universe's scale, the complexities are represented by the stars and solar systems, which all evolve toward an imminent collapse. The dissipation (disorder) is a general tendency in the universe, and it was established at the Big Bang.

The ***dissipation represents the collapse of the proposed solutions or unanswered answers***. The path in search of the ultimate answer has to bifurcate very many times until the system's coherence turns lost, while no answer can be reached.

In my opinion, the black holes, representing immense attractors, which tend to absorb all dissipated elements of the collapsing star systems.

Are they recreating what was apparently lost in dissipation?

* * *

Human civilization, as a manifestation of the phenomenon of augmentation, is not the first such event on our planet.

Without entering any details, I would consider that the ***era of Carboniferous***, developing from cca. 360 to 300 million years ago, represented such a large manifestation (with a ***duration of 60 million years***) of ***augmentation that affected most of the plant species***. The only continental mass at that time was the supercontinent Pangea. Some 50 million years after the end of Carboniferous, during the ***Mesozoic era*** (248 to 65 million years ago) begins the manifestation of a second large phenomenon of augmentation that, this time, affected the population of animals. During its ***150 million years duration*** occurred several natural disasters (during Triassic that lasted from 248 to 206 million years ago), which wiped out up to 90 percent of all species.

During Mesozoic lived the *dinosaurs, which have been the* ***expression of augmentation manifested among the animals.***

In both cases, of Carboniferous and Mesozoic, it occurred an augmentation among all species, first in the plant, secondly in animals. There is of little importance the fact that in both cases, it existed a significant growth in the size of plants and animals.

I like to say that if one details the geologic and biologic history of this planet, there is possible to identify several more episodes of augmentation. But maybe none of them have the ***ample significance of Carboniferous, Mesozoic, and Anthropocene***.

* * *

I like to talk a bit about the alleged stages of extinction manifested during Carboniferous and Mesozoic eras.

During Carboniferous era occurred a minor extinction event placed around 305 million years ago. As it was documented, "it altered the coal forest that covered the equatorial region by fragmenting it into isolated islands. It caused dwarfism and extinction of many plants and animal species" (Wikipedia).

"A mass-extinction marked the beginning and end of the Mesozoic Era. The event that caused the transition from the Paleozoic era to the Mesozoic era was the greatest extinction this earth has seen. The era of Mesozoic ended with "the Great Extinction," which marked the end of dinosaurs" (Wikipedia).

There is a trend to explain such "extinctions" by "natural catastrophes."

I am far of denying the possibility and occurrence of natural catastrophes, but what I like to enlighten here is the aspect of collapse that naturally manifests within processes of augmentation.

* * *

As far as one would agree that here, we deal with processes of large augmentation, one must admit that the "phenomenon of augmentation typically ends through a process of collapse."

The scale of augmentation dictates the scale of the collapse.

In human history, most of the empires ended through collapse. Greater was the empire; bigger was its collapse. In all such collapses, no natural catastrophes were explicitly implied.

In the meantime, there is enough evidence that natural catastrophes have occurred in those eras. Obviously, some exceptions can be documented, but they will not affect the general rule.

The topic of the collapse of civilization created a very controversial debate in literature and media for many years. Hence, there is no need to resurrect this subject.

My intention is only to bring a very distinct view on this same issue that is the view associated with augmentation phenomenon.

People always like to find excuses for everything that goes bad, but in most cases, here are ignored or unknown natural rules, which manifest synchronously, and make alleged coincidences to appear everywhere.

* * *

The process of augmentation produced by the human mind is fundamentally distinct that its natural occurrence.

It originates in language, but *in the brain, the neural structures in charge with the development of language are associated with structures responsible for the control and planning*. Hence, ***these three approaches evolved together***.

However, language itself introduced "linearity" in thinking. The communication component developed by language implied increased "socialization." Thinking-as-talking was the first new attribute being introduced to speakers' thinking.

The language developed in the left hemisphere, where also exist the structures for pattern separation, which structures gradually evolved in the capacity for ***synthesis-analysis*** that generated the ***logic thinking***.

The new attribute of logic thinking introduced ***quantitative thinking*** as opposed to previous qualitative thinking. This led to ***mental computations and mathematics***.

As one can see, because of language evolution, the human mind adopted approaches, which contrast the previous way of functioning.

It comes linearity vs. nonlinearity, quantity vs. quality, logic vs. correlations, logic thinking vs. emotional thinking, verbal vs. visual, and ***linear understanding vs. nonlinear/ correlative understanding***. Several other attributes can be added to this list of mental changes, while a very important one was the introduction of a type of ***habitat-thinking*** vs. an environmental-thinking. Domestication of animals and plants was an expression of control-thinking and planning-thinking. Construction of habitats also changed the previous naturalist-thinking dramatically.

All previous mental attributes were created by natural evolution. ***The introduction of the language produced new attributes, which oppose the previous but natural ones. Altogether, the new attributes were bounded to produce artificial behaviors***.

Maybe the most disruptive approach was the creation of artificial computation that replaces mental computation. A similar significant disruption was (starting two hundred years ago) the introduction of machinery and industry, which both are intended to replace human physical effort and to provide mass production of goods. All these are a manifestation of augmentation.

Hence, the mentioned ***new but artificial attributes represent the new mind capable of producing an "augmentation of artificial approaches."***

By now, it becomes obvious why human augmentation has evolved so distinct than natural augmentation phenomenon. Even then, the way such artificial augmentation behaves it mimics, in general, the rules of a natural phenomenon because here are universal rules, which cannot be avoided.

Besides the enormous expansion that guides human augmentation, and besides all adaptations produced in the human brain, even today, humans are biologically unchanged compared to our Late Paleolithic ancestors.

While human augmentation started thirty thousand years ago with language advances, it matured and begun to manifest only in the last ten thousand years. However, the building up of our civilization only occurred in the last five thousand.

We have created a technological environment that continues to produce an immensity of structured and unstructured data that increases on an ever-growing exponential path. All data produced by such environment is fundamentally artificial.

Here, the **unchange**d **human biology is** stressed **by the *immensity of data"*** generated by augmentation.

To conclude, augmentation affects and models the behavior of the system. It generates, to a system, the tendency to grow on its macroscopic scales, because it is a macroscopic phenomenon.

COULD WE TALK ABOUT TRANSMUTATIONS IN HUMAN BRAIN PROCESSING?

As assumed, perception is an "interpretation" of sensory information. Hence, it is not that direct physical and raw translation of environmental information into our brain, as many people tend to think.

By this assumption that contains very many scientific proofs attached, one may say that perception is not objective but subjective. And this is true. The brain perceptual system attempts to make sense of the input.

Now, the first important question to be answered here regards the nature of the input.

Structuring Data.

Hence, I have to quote Dr. Vladik Kreinovich book dedicated to the anniversary of Professor Boris Kovalerchuk and published by Springer in 2017. It refers to the role of uncertainty played in perception.

"Events that could not happen by chance are immediately perceived. Meaningful features and interesting events appear as large deviations from randomness. From sequential and unstructured data, we derive a qualitative measure of such deviations."

As the author has explained in the above paragraph, in this mental process, "intervenes a qualitative measure." Thus, it is not a quantitative computation or analysis.

"Unstructured is data without an explicit data model, but with some internal geometrical structure. Data comes packaged into objects."

The author implies the presence of "unstructured data" that has only a type of internal geometrical structure.

Now, Dr. Kennedy Elliott in chapter 3 of his book, "Human Perception and Information Processing," specifies that "perception is the process by which we interpret the world around us, forming a mental interpretation of the environment. The brain makes assumptions about the world to overcome the inherent ambiguity in all sensory data, and response to the task at hand."

Dr. K. Elliott sees this mental processing as an **interpretation** of raw input of environmental data.

According to Dr. Vladik Kreinovich, the unstructured data, that represents the maximum entropy provided by the environment, contains an immense amount of "uncertainty."

During evolution, all brains, not only the human brain, turned to make a selection within their "unstructured environmental input," and they eliminated as much as possible the random inputs. So, their perception was adjusted to be the first stage that operated the selection.

As many scientists have proved in their research, we see what we want to see not what it is out there.

I would say, perception evolved toward selection in all brains by reducing the environmental "uncertainty."

But this reduction process affected only "data." It selected among input data some that make sense for each lifeline. It selected "orderly" (with as little randomness as possible) aspects, which seem to be cyclic or circular, can be predicted, and therefore, controlled.

The evolution drove the perception of the unstructured data into semi-structured data that enables some "grouping" and "hierarchical approaches."

As the specialists estimate, such semi-structured data may amount to 5–10 percent of all unstructured-structured data combined. But here, as the volume of data decreases, its relevance increases.

As Zat Rana explains in his October 20, 2018, paper, "The Two Types of Knowledge, "You create mental concepts in your mind about what is important and what is not, and then, these concepts shape your future perceptions by using the already-selected knowledge to further select knowledge. Whether you are learning to play a sport or simply trying to create a more accurate mental model of reality in your mind, you are working with a variety of experiences, and within those experiences, you have to choose and reinforce the ones that are the most useful to you. In this way, everything you do is essentially an experiment that gets refined and corrected with experience and practice."

It is my opinion that in human development the language played a very large role in forcing unstructured data gaining structure, because the use of words and groups of words, could provide means for classification and analysis-synthesis. The use of language changed human thinking that became sequential, produced a thinking-as-talking, allowed the development of analysis-synthesis (or logical-thinking), and generated imaginary-thinking.

Let's return to the main topic of this paper that is: a chained transformation concerning data-information-knowledge, and how this relationship develops or is accomplished.

Information is data put in context, and this process succeeds when it could add a "meaning." The content of information is like a message because it must explicate "something" about "a particular something," and that "message" processing identifies a "meaning." We associate that "explication-message" (that is a description in itself) with a particular "explicate-identity."

In sum, the unstructured data has inner geometry, but the description is too vague. The information has a clear description. The distinction between the two resides in their amount of "uncertainty." Selecting "ordered descriptions" was a process that occurred in all brains, it increased the "certainty," allowing some prediction and control, which both were essential for surviving. Every brain creates in its selecting processes a world of "order," where that being can make a survival.

Here, data is abstract, and information defines such abstract in its message. A description fundamentally remains abstract.

Knowledge is neither information nor data. It signifies things we know about as "facts." It abstracts a direct or indirect experience. But its definition suggests that knowledge derives from information and information from data. Hence, knowledge has also an inner geometry.

As mentioned, the brain, as it evolved, it selected the inner geometry that repeats itself in various data. This repetition of geometry eventually built a geometric environment, where the items were repeating themselves.

Such a new frame of the mind allowed to predict such repeatability, but also helped to find "patterns" as repeating shapes and events.

The geometric environment was subject to emotional thinking that defined the geometry by its **qualities** derived from experience. It was beautiful or ugly, tall and short, fast and slow, dangerous and favorable, good and bad, and so on.

Every such definition was associated with some **inner geometry**. Human confrontation with an existential danger (in my opinion) was advantaged by a chain of mutations, which produced a closer relationship between individuals using better communication. The same chain provided new means for a higher social organization, for an improved planning capability.

There is known that language, planning, and control are processed by the same structures of the brain. Thus, language evolved simultaneously with the other mentioned skills. Some say, talking is, in fact, an attempt to extend control over the listener.

Language allowed the spread of innovation due to communication, but also it gradually replaced ancestral nonlinear thinking with sequential, linear thinking favored by thinking-as-talking.

Nonlinear thinking, as an evolutionary result of vision being paralleled by other four senses, was favoring a type of parallel computation in the brain that was summarized by what we call "correlation." The interpretation of such correlations produced "understanding."

Because of language evolution, emotional thinking was gradually replaced by a highly structured logic thinking that was capable of analyzing and synthesizing, while gradually inserting the limitation of sequential processing.

This change toward more structured data also changed the thinking system from nonlinear into linear.

Linearity brought a mental connection of inner geometries of all things, producing the continuum of the space. It has produced a unified by geometry environment.

Domestication of animals and plants occurred in the era of qualitative and emotional thinking. Some scientists consider that the domestication process was, in fact, an extension into the environment of new concepts about social relationships, and where the environment was intimately embedded in society. Complimentarily, new social relationship provided a means for processes of self-domestication among human groups.

The practice of agriculture introduced a new need for quantitative-thinking. Gradually the quantity prevailed over the quality in general thinking.

One could say that the numbers originate in the same inner geometry inserted in every piece of data, while the mathematics simply express the inherited geometric background.

Maybe the abstract existed in the human brain, but it became expressed in the form of art and music only when it turned semi-structured. It records the transformation of unstructured into information and meaning.

The art revolution in South Africa some seventy-five thousand years ago was the first stage of the process mentioned above. The second stage, occurring forty to twenty-five thousand years ago, reflects the generalization of a transformation of unstructured into semi-structured and achieves a scaling up of informationalization defining the structured.

The Role of Language In Structuring Information.

It is assumed that sequential structure would be fundamental to the comprehension of language.

Languages differ from each other in various ways, like by sentence structure (syntax), by world structure (morphology), by sound structure (phonology), and by vocabulary (lexicon).

Al these are structures, and the human mind has structured them from unstructured forms of sound. It is assumed that a computational cognitive system generated the language structures that are related to grammar. General cognitive processing interacts with linguistic rules, which are structure-building. This process produces belief information such as memory, attention, reasoning, and several other inference processes.

I will quote Malte Zimmermann, who wrote in Oxford Bibliographies on July 27, 2016, the following explanation:

> "Information structure plays a crucial role in embedding linguistic utterances in ever-changing

communication settings and contexts, and it does so by imposing structure on how information conveyed by an utterance is linguistically expressed."

"Information structure is responsible for an efficient information transfer between interlocutors, where information transfer consists in the updating of interlocutors mental models of the world, and in the establishment of mutually shared knowledge bases (common grounds) through the exchange of linguistic utterances. Linguistically coded information is structured in a specific way, such that it fits the context of utterance and knowledge states of the discourse participants. The information structure of linguistic utterances is typically reflected in their grammatical form. Linguistic marking of information structure facilitates information to update and the actualization of belief states".

"Information structure refers to the concrete structural realization of information structure categories in linguistic utterances. The information structure categories themselves are taken to be universal".

Information structure could be realized by a variety of mechanisms, like intonation that modifies the pitch, or specialized focus (affixes, discourse particles, contrastive particles, etc.). It also intervenes a hierarchical structure that plays a key role in most linguistic descriptions.

Until now, I talked a lot about rules, which all would be attributed to grammar.

I have now to quote another paper, "The Emergence of Grammar: Systematic Structure in a New Language," written by Wendy Sadler, Irit Meir, Carol Padden, and Mark Aronoff and published in PNAS February 15, 2005. The authors make a "linguistic description of a language created spontaneously without any apparent external influence in a stable existing community." They "describe the syntactic structure of Al-Sayyid Bedouin Sign Language, a language that has arisen in the last seventy years in an isolated endogamous community with a high incidence of nonsyndromic, generally recessive, profound prelingual neurosensory deafness."

> "The Al-Sayyid Bedouin group was founded cca. two hundred years ago in the Negev region of present-day Israel. The group is now in its seventh generation and contains cca. three thousand five hundred members, all of whom reside together in a single community exclusive of others. Consanguineous marriages have been the norm in the group since its third generation. Such marriage patterns are common in the area and lead to very strong group-internal bounds and group-external exclusion".
>
> "Within the past three generations cca 150 individuals with congenital deafness have been born into the community, all of them were descendants of two of the founders' five sons".

Now, I have to make several assumptions. First, the community said to live in isolation has seven generations in two hundred years. This means that one generation in this particular case represents thirty years. A couple of founders had five male born children. Here, it must be assumed that those five sons brought from outside into their community five wives. This makes the second generation.

As the authors of the study described, from the third generation, meaning the children of those five sons and their wives, it started a consanguineous chain of marriages that went now in the fourth generation of such consanguineous marriages.

As the study finds, now the community has 3,500 members from whom 150 are born deaf. It is to assume that the consanguineal marriages are the reason for born deaf children. But here, the authors specify that the born deaf individuals represent the third generation of such manifestation. It means no born deaf children occurred during three generation since the community was founded.

The rate of deafness is now probably close to 15-20% of the entire community, considering all three past generations.

The authors describe, "All deaf individuals show a profound prelingual neurosensory hearing loss at all frequencies, have otherwise normal phenotype and are of normal intelligence. Scott et al. identify the deafness as (recessive) DFNB1 and show that it has a locus on chromosome 13q12 similar to the locus of several other forms of nonsyndromic deafness."

"The deaf members of the community are fully integrated into its social structure and are not shunned or stigmatized. Both male and female deaf members of the community marry always hearing individuals."

I am not sure that I am right in these two assumptions. Further on, the authors describe this community as practicing a sign language that is also common among some other neighboring while also isolated communities from the Negev.

The study researched the rising of the new language spoken by this community that is a new language that seems to be unrelated to any other known language.

It is the study about how a new language has appeared from a sign language in conditions of isolation.

"Kirsch was first to report that the deaf members of the community and a significant fraction of its hearing members communicate using a sign language. Siblings and children of deaf individuals and other members of a household often become fluent signers. Members of the community generally recognize the sign language as a second language of the village. Signers readily use their language to relate information, such as descriptions of folk remedies and cultural traditions, some of which are no longer in force".

"The **statistical study of the sign language of this community show significant regularities in the order of signs in the language**. This word order cannot be attributed to the ambient spoken languages (like Arabic or Israeli). Such a particular word order is not characteristic for any other language. The robust word-order pattern exhibited by the data is all the more striking because it cannot be attributed to the influence of other languages".

"Susan Goldin-Meadow finds **consistent word order appearing early in their gestural productions**. Although these children have no contact with other deaf signers, second-generation ABSL signers consistently produce sentences in which predicates appear in the final position, but unlike the home signers, they produce longer sentences with two or more nominals appearing before predicate".

This study indicates the **paramount role of sign languages, which had already developed their own grammar embedded**

into them. Hence, the presence of grammar presupposes a structuring of data in the human mind and a sequential arrangement of meanings.

It also indicates that a sign language could be quite fast developed into uttering if the vocal development occurred.

The authors conclude, "Of greater significance to us than any particular word order is the **discovery that very early in the life history of a language**, a conventional pattern emerges for **relating action and events to the entities that perform and are affected by them, a pattern rooted in the basic syntactic notions of subject, object, and verb or predicate**. Such conventionalization has the effect of liberating the language from its context or from relying on the semantic relationships between the verb and its arguments. Once the language has had time to accrue such mechanisms as verb agreement, making properties of subject or object, or **case making on the noun to indicate their relationship to the verb, the roles of participants can be made clear**, even without consistent word order. It is a rare verification of the unique proclivity of the human mind for structuring a communication system along the grammatical line." I have to explicate one important aspect: two hundred years ago, when the above study dates the origin of the described process, humans in the Middle East have long ago acquired all their mutations, anatomical and neural, which are considered fundamental in the development of language.

Hence, those people living in isolation in the Negev desert were equipped with full lingual producing capabilities, but their isolation prevented them from their full use.

Contrasting with Negev case, there are the aboriginals of Australia-New Guinea-Oceania, and Amazonia-Americas, which can be defined as zones which fall outside the major influence of the geophysical events with probable mutational outcome in the last 40,000 years to 2,500 years.

Data, Information, Meaning, and Knowledge.

It can be assumed that constant while gradual structuring of data contributed to enlarge the conscious processes. Eventually, the semi-structured data shaped the preconscious and unconscious processes.

Mineral Intelligence Capacity Analysis (MICA) Organization indicates that "research in knowledge management would suggest that only 15 percent to 20 percent of valuable knowledge has been captured and is available through a tangible form. The need, therefore, for systems that mobilize tacit knowledge to create explicit knowledge is important" (Dalkir, 2005).

MICA study (Deliverable D3.3) continues,

> "Knowledge conveys expectations, instructions, and rules, and it often provides answers to 'how-to-question' (Ackoff, 1989 and Canals, 2004). **Knowledge is closer to the action than data and information**, and this is the main reason why it is seen as valuable. Experience, truth, judgment, and guidance provision are key components of knowledge and are interlinked in complex ways (Devenport and Prusak, 1998). **Data and information are visual elements, whereas knowledge is something we know, and it requires analytical thinking**."

> "Wisdom is a much more complex term to define; it is multi-dimensional and requires diverse knowledge. **Wisdom often represents the overall picture** (Barlas et al., 2005). **It combines knowledge to formulate actions and achieve goals**; it increases effectiveness".

> "Spatial data represents points, polygons, lines, or pixels They include a location, shape, size and orientation and they are therefore multi-dimensional".
>
> "Non-spatial data relate to a specific, defined location. Statistical data are often found in this category, but text, images, and multimedia data are also part of this group".
>
> "Information represents structured data. The role of information is to inform someone about something and to satisfy an information need.
>
> Very often, data are transformed into information through model development".

For example, words and sentences have parts that combine in patterns, showing the grammar of the language.

"**Knowledge is contextualized information**, **and it is produced by combining available information with expertise, insight, and intuition**."

"**Intelligence is associated with an action taken based on the knowledge available**."

Transmutation of Data, Information, and Entropy Behavior.

As most scientists agree, information is a fundamental property of the universe. But according to Lindauer's principle states that **it takes energy to perform irreversible transformations on information**, but it cannot be concluded that energy and information can be converted into one another.

Everything is fine, but here how data transforms from something into something else?

I picked up from online Data Science Central a blog posted by Mkhuseli Mithuwane on September 29, 2015, titled "Data Transmutation vs. Data Analytics: Round 1." I found into this blog some inspiration about how nature succeeds to transform and transmute data without human interference. There is to assume that this natural mechanism works in all brains, human brain included.

> "**Data Transmutation creates a synthesis of multiple events that are first encapsulated into a single expression, then transposed into a math function and finally transmuted into a coherent data point**.
>
> It condenses Data into mathematically calculable numerals and symbolic expressions that weight the factual permutations of events and occurrences. This distillation of Large Data sets into highly concentrated but rational Data points is very helpful. Data transmutation is a bit like gene extraction. All Data tells a story, and every story has a core phenotypical structure and genome. ***Data transmutation*** **is a way of delineating the Data into DNA strands that** ***represent the foundational archetype of the story the Data is telling***".
>
> "***Transmutations mine the primary consequence of events and occurrences*** **by isolating the systemic functions of Data; thereby** ***extracting only salient (important) truths*****."**
>
> "When you Extract, Transform, and Load Data, you are essentially **taking a series of events and fragmenting them into scalable features**

> (for Algorithmic inquiry). This fragmentation increases density, widens factual parameters, increases variation, and ultimately diffuses the efficacy of the story".
>
> "However, ***when Data is transmuted*** **into a condensed form, factual parameters aren't unnecessarily expended, the *features become more salient(important)*, *the density remains the same* and *variation is kept at a healthy level*.**"

Dr. M. Mithuwane makes sense of Data processing that results from Extract-Transform-Load and compares it with Data Transmutation.

In nature, the processing is eventually distinct than in mental computation, because in the first one we introduce an "algorithm" that has the role to model the unstructured and to give it a structure that is a manmade structure.

However, we know what all brains do: they filter the unstructured data by choosing to retain the **data** with the slightest (smallest) variation **that can stand with little or no alteration against the surrounding randomness** (noise). Such no-variation Data is processed by consciousness.

Overall, the brains perform a type of transmutation in many respects, similar to the description given by Dr. M. Mithuwane.

As I presented in several above paragraphs, most researchers consider that "unstructured" data must have an inner geometry. Thus, I would assume that the **brains should compute such type of geometries without relying on any digits and numerals**. And therefore, I would assume that the **natural transmutation should alter those geometries, transforming one type of geometry into another**.

I think that the **Paleolithic type of qualitative-thinking** or thinking based on qualities **was a geometric type of thinking**, which was investigating and comparing various inner geometries displayed by the subject residing in the environments. It was a qualitative type of thinking, and it was inherited from the evolution mechanism of natural processes.

The whole natural idea of generating "negentropy" was to "inform meaning." "**Negentropy" follows the principles of information that produce causation and meaning**.

E. G. D. Cohen and R. L. Merino in a paper titled Clausius' Entropy Revised, published on January 10, 2014, explain:

> "We argue that the main feature of such systems is not the entropy production, but the organization of the flowing currents in such systems. These currents do not only have dissipation of entropy production but must also have an order or organization needed to flow in a certain direction. It is the latter, which is the source of the equilibrium entropy when the fluid goes from a local equilibrium state to an equilibrium state. This *implies a transmutation of the local equilibrium currents' organization into the equilibrium entropy*. Alternatively, ***when fluid goes from an equilibrium state to a local equilibrium state, its entropy transmutes into the organization of the currents of that state***".

Undoubtedly, the quotes, from the abstract of this mentioned paper, refer to physical processes, where take place a **transmutation of equilibria states that define the entropy of a system**. And this is the source of the equilibrium entropy. There is no doubt that within the biophysics of brains, similar processes take place among the currents of the fluid systems, which flow in the brain.

As assumed by the authors of the above paper, the *transmutation of the equilibrium states causes a **transmutation of entropy** in those systems or any other similar systems.*

Within my assumption, such systems could also be neuronal in nature. And then, this theory could disclose the ***entropy transmutation mechanism in brains***, human brains included.

Frank Stein, manager at IBM, in his blog Information Alchemy: The Transmutation of bits and bytes into Knowledge, published on March 6, 2013, explain:

> "Today, in the field of information science, we talk about **Information Alchemy, converting data into information, and then into knowledge**. Data is defined as the raw characters or numbers, whereas information is defined as the processing of data into various relationships, so they have some meaning. Dr. Eisenberg at the University of Washington describes **knowledge as the "collected, combined, organized, processed information for a purpose**."
>
> "Today, the transformation of information into knowledge is primarily done in people's heads. The time available for people to analyze information to gain insights (knowledge) is the limiting factor in the production of new knowledge today. To say this another way, **we are now information-rich, but knowledge-poor**".

To conclude, I made several assumptions here regarding the possibility that ***all brains, including the human brain, have the eventual or theoretical ability to produce some forms of transmutation***. Such a phenomenon can encompass processes transforming data into information and knowledge, but also it may transmute entropy, or simply it may affect the workings of various fluids flowing in the neural systems.

As it seems, the idea that the structuring data and information requires a certain deal of energy opens the question about the role played by increases in the energy of the brain rhythms. I assumed that during the geomagnetic excursions it occurs an increase in the energy provided to the brain rhythmic activity.

Hence, **an increased energy of the brain rhythms seems to be the vehicle that produces increased structuring of data and information**. It helps produce a better and richer expressed language. Here, as it is known, many mental illnesses display a reduction of rhythms' energy, making the subjects to lose some of the ability to structure data and information. It produces a language deficiency. However, **this issue of information-transmutation that could influence the behavior of data**, if documented, may imply a paramount **role being played in producing the phenomenon of consciousness**.

HYPOTHESIS OF HOMO LOQUENS

In several papers, I investigated the changes in biological fundaments that modeled the human brain, generating the current "modern brain."

During this large investigation, in the end, it turned significant to me that two factors produced the most major changes.

In my opinion, the main factor that gradually changed the hominin brain was the repeated occurrence of geomagnetic events, which favored the penetration of cosmogenic radiation through the Earth's atmosphere and which transformed into an atmospheric increase of C14 isotopes. As it seems to be the case, the effect of absorbed C14 isotopes regulate the production and/or inhibition of nitric oxides. A high concentration of atmospheric C14 isotope is demonstrated that, during its assimilation in plants by photosynthesis, it produces sucrose that is 50% glucose and 50% fructose. Because of the presence of C14 isotope, the product is C14-glucose. The animal organisms assimilate this last product during feeding on vegetables. However, **once C14-glucose is assimilated in the animal organism, it inhibits nitric oxide production**. The experiments show that **when a high C14-glucose is present in the biological system, it decreases the release of nitric oxide (NO)**. Here, the nitric oxide becomes involved in transporting and dispersing of glucose in the entire organism; hence, **nitric oxide decreases in the brain and nervous system because it increases someplace else, like at the level of the muscle fibers**.

The nitric oxide basic effect is vasodilation and favor inflammation of tissues. But, when the neurotransmitters are involved, it favors communication between the cells and between the neurons.

As has been observed, during various experiments on plants and animals, **a reduction in the released nitric oxide causes the opposite effect that is vasoconstriction**.

As a team of researchers from Nanjing University (China) demonstrated in their paper "*Bidirectional Regulation of Neurogenesis by Neuronal Nitric Oxide Synthase Derived from Neurons and Neural Stem Cells*" (published in Stem Cells in 2010), "the neuronal nitric oxide synthase (nNOS) negatively regulates adult neurogenesis."

Another paper ("*Nitric oxide negatively regulates mammalian adult neurogenesis*") of a team of researchers from Cold Spring Harbor Laboratory, NY (published August 5, 2003, in PNAS), comes to support the same idea. These authors say:

"Here, we report that **nitric oxide (NO) acts as an important negative regulator of cell proliferation in the adult mammalian brain**."

The **vasoconstriction** increases the tension inside the blood vessels, **producing a hyperbaric effect**.

Two other papers ("*Long Course Hyperbaric Oxygen Stimulates Neurogenesis and Attenuates Inflammation after Ischemic Stroke*," written by a team of researchers from two Taiwan universities, and "*Hyperbaric oxygen therapy promotes neurogenesis*," written by a team of researchers from Loma Linda University, US, and Chongqing Medical University, China) demonstrate how **hyperbaric procedure that involves vasoconstriction effect stimulates neurogenesis**.

Thus, I would say that now I have sufficient arguments to support a **hypothesis about the direct and indirect role of geophysical events, named geomagnetic excursions, together with their cosmogenic radiative input**, which both have changed the C14 isotope concentration in our atmosphere, while **that resulted in a fundamental influence on the brain mechanisms**, especially for the process of neurogenesis, **in hominin and humans**

As the literature indicates, humans present a different pattern of adult hippocampal neurogenesis as compared to other mammalians. In humans, the vast majority (90%) of neurons in the dentate gyrus are subject to exchange, compared to 10-30% in other mammalians investigated. Humans show a less pronounced age-dependent decline during adulthood compared to mice and other mammalians.

As it seems, the negligible addition of new neurons to the olfactory bulb (OB) after the perinatal period is unique to humans. As various studies show, in other mammals, all lifelong, the neurons generated in the subventricular zone are integrated preponderantly in OB (olfactory bulb). But in humans, after birth, not such a process exists. It indicates that human neurogenesis takes a distinct path compared to most mammals.

Here, there is a question: why such a distinction arises? I would say that life-forms' evolution is not linear as we generally tend to see it. In the case of humans, we consider an entire range of species and subspecies, which originate from monkeys as our evolutionary connections. I would not debate this approach, but I would consider a nonlinear view, where the fundamental argument is related to neurogenesis.

As one could observe, the way neurogenesis evolved in mammals and primates points to a moment when a clear separation singled out the humans against the rest. In the meantime, if one counts the subspecies and species evolved out of Australopithecus, in the last 4 million years since the first one originated, one finds a plethora of species and subspecies. I would say that the archaeology probably was not able to find more than a few of all species, which existed in this large family of so-called hominins.

Why developed such a huge family of hominins? The first probability to consider refers to somehow the high-sensibility of this hominin family to environmental changes. The second probability regards a large number of varied or distinct species and subspecies from where it would be chosen natural optimality.

Within a nonlinear perspective, here should exist a certain causality that took advantage of an existing biological sensitivity. As mentioned, the humanoid neurogenesis was affected by a particular geomagnetic occurrence that I called geomagnetic excursions. Let's have listed the main geomagnetic excursions of the last 300,000 years:

- Calabrian Ridge 260 kyr;
- Pringle Falls 211 kyr;
- Iceland Basin 188 kyr;
- Blake 120 kyr;
- Norwegian-Greenland Sea 80-70kyr;
- Laschamp 40kyr;
- Mono Lake 33 kyr.

Regarding *genus Homo*, it has its own family branches.

It started with Homo habilis around 2 million years ago.

The most important member, before us, was *Homo erectus* that appeared 1.8 million years ago as the fourth branching of Homo family. It mostly disappeared 300,000 years ago, but it is known that locally it survived until recent times. It evolved in at least 12 distinct subspecies.

Next would be *Homo antecessor* (800,000 years ago), *Homo heidelbergensis* (800,000 y.a.), *Homo naledi* (300,000 y.a.), *Homo neanderthalensis* (200,000), *Denisovans* (they split 236-190 kyr ago).

Here comes our subgenus or *Homo sapiens*. The earliest is dated around 190,000 years ago as an archaic form. Around 120,000 y.a. appeared the true sapiens.

In my opinion, the last member of the subgenus Homo sapiens is us; **we originated as *a speaking-subspecies or even a new species* around 40,000-30,000 y.a.**

As it seems, the **clue of such complex** but somehow contradictory **evolution is the nitric oxide** (NO). It is a gaseous molecule synthesized by the enzyme nitric oxide synthase (NOS). It acts as a neurotransmitter. Also, it is a part of the signaling pathways that operate cerebral vessels, neurons, and glial cells.

There is known that NOS exists in three isoforms: endothelial nitric oxide synthase (eNOS), neuronal nitric oxide synthase (nNOS), and inducible nitric oxide synthase (iNOS).

Each of these isoforms influences the brain in distinct ways because they are active in different sites within the brain. The NO signaling pathways regulate the cerebral blood flow (CBF) during rest and various stresses.

In the meantime, there is **a growing body of evidence that NO plays a significant role in very many biological processes**. It is an important mediator in the central nervous system, and would be implicated in a wide variety of neurological diseases. For example, in degenerative brain processes, which affect the CNS, the nNOS level rises rapidly. In some pathological conditions, as inflammations, the iNOS level is very high.

Overall, nitric oxide is seen to act as a "double-edge-sword."

The **role of nitric oxide could contribute to either cell survival and cell death**. The ultimate answer is the untimely transfer of nitric oxide from one protein to another. There is evidence that NO transfer could simply lead to cellular suicide.

NO can act as a modulator of the inhibitory processes within the synaptic transmission. But also, it can modulate the activity of ion channels and regulate the ion exchanges.

The C14-Glucose, ingested through hominin and human feeding, is known to have, in general, an inhibitory role on nitric oxide. It would be very difficult to assess a certain brain organization at a particular time, or what is the exact effect that occurred.

However, within my hypothesis, **a large rate of rising and disappearing (extinction) of subspecies and species of hominin and humans could be largely correlated to a recorded variety of geophysical events, which all produced short and/ or long episodes of high atmospheric concentration of C14 isotope**.

Now, I like to refer to Homo sapiens evolution. As it is known, it evolved from an "archaic type" (around 190ka) into an "incipient sapiens" (120ka). Nevertheless, around 80ka-70ka, it appears the first manifestation of art in South Africa. It is true also that Neanderthals seem to enter a phase of high development during the same epoch. But 42ka, a chain of strong geophysical events started to unfold. This chain generated a revolution in language for Homo sapiens but led to the extinction of Neanderthals and Denisovans, while the last surviving Homo erectus and other hominins also reached their extinction phase. It is a critical point, where **Homo sapiens encountered fundamental changes**, like the change in skull geometry toward rounding, the conjugates switched the brain lateralization, dual-use of the brain neural circuits (visual + verbal), some new neural circuits (for verbal only), changes in the neuronal processing speed (speed diminishes), while the entire body suffered a process of gracilization (as an extended form of neoteny). **The change in the skull geometry is seen as fundamental for a very distinct development of language. It had put the language-related sections in a central position, and by rounding, it diminishes some circuitry distances**. A switching of correlates had paralleled the other developments in the brain while changing the pattern of lateralization.

Because of such, as mentioned, but fundamental outcomes, my hypothesis proposes here that **it appeared a new species**. **I would call this new species *Homo loquens*.** As the archaeology indicates, the rounding of the skull appeared some 30ka but was fully completed during the Holocene optimum (cca. 8ka).

The language was generating anatomical changes because it was a technology that became embedded in our biology. It was a full biological response that needed to be developed by this new species: from genetic changes to new genetic expressions to anatomical reshaping.

The language was and is a technology in itself. It can produce an infinity of other technologies by playing with its assembling/disassembling capacity.

Time and again, the language came with a set of features, where communication was one of them but not the most important. The other features were planning, control, and socialization. Over time, the importance of each feature changed. Planning implies creating a particular type of language. Hence, planning generates all other languages, including those we call algorithms. We can create an infinity of languages and algorithms.

Control confines us to a system of minimal variability that excludes the rest as unperceivable and unthinkable. Over time, we have learned how to deal a little bit better with variability. But even then, our progress was minimal. We still hope that the features being developed by the Artificial Intelligence and Quantum Mechanics will propel us further in understanding and dealing with uncertainty and its variability. Right now, it is only a promise.

The feature of socialization seems to be one of the best developed.

The advent of quantum mechanics would have a hard to guess result because it would introduce new features.

On the other hand, the Artificial Intelligence tends to multiply all four features. But it would also remove some traditional advantages we have developed and exploited in time.

A recent study of Harvard Medical School and its Psychology Department, performed by Elinor Amit and Evelina Fedorenko (published on May 11, 2017, in The Harvard Gazette, and journal NeuroImage) was able to disclose several fundamental details of modern thinking. The authors revealed that people use inherited **visual thinking about those things which are closer to them**. In the meantime, people use **inner speech**, as their new verbal ability, **when they contemplate far-off things**. In short, people use visual thinking to deal with self, in-group, past, present and near-present. By contrast, verbal thinking is used when dealing with somebody else, or an outgroup, or future. The researchers found out that visual deals with past and present, and verbal with the future. Here, **there is a clear indication that the development of verbal thinking helped us to define the concept of the future**.

But more important, it appears that **language helped one individual deal with another**, while ultimately, it helped to define and promote the self.

It also helped the individual to become open to out of group relationships and contacts. A less talked aspect is **the relationship that started to develop with the environment**. Such a new approach turned paramount in human evolution: it led to the domestication of animals and plants.

In sum, **all this changed the type of social relationships, and it has led to a change in the human diet. All became promoted only due to language**.

In the meantime, the researchers found out that, at present, there is no pure visual or pure verbal thinking because of both turns mixed in every expression. Hence, the **recently developed verbal mode is**, by now, **very well implemented in all of our brain circuits**, contributing to a certain part of every expression.

Even then, one of these two modes of thinking appear to be predominant in one expression or another.

As a negative outcome, experimentally, it is demonstrated that those people very good at verbal thinking score about average, or even low, on the object and spatial visualization tests.

It is a clear indication that **almost all old visual circuits admit the interference of newly developed verbal mode**. This is an interesting mental approach because **each of these two modes shows a very different speed of processing** (it is admitted that visual is at least 10,000 times faster than verbal).

Overall, this Harvard study provides valuable information about **how human thinking radically changed since the advent of language**, which occurred starting 30,000 years ago.

I see that **the most consequential result of verbal was the development of "*imaginary thinking*" accompanied by its "*factual*" products**. This "***imaginary***" mode exploited the verbal capabilities to a maximal extent that implies an "***infinity***" of mental manipulations, which generate "factual" products.

However, the **language was not a human invention**, as I repeatedly misspelled before. **It resulted from a complex play of evolution that adapted the hominin and human biology to induced while particular geophysical constraints**.

At the time of its revolutionary development, language helped us prevent an extinction. By contrast, all other species of hominins disappeared (including the Homo sapiens sapiens who were replaced by our new species), not coincidently when humans began to turn into a new species cca. 30ka.

How much time this change of faith bought us? It may sound awkward when I say that we have replaced Homo sapiens sapiens in that process that created a new species that is us.

In support of my opinion, I like to add here a controversial idea that still makes sense. The Bible says: "**In the beginning was the *word*.**" Then, I ponder: who spoke those *words*? The

Bible has no mentioning of any "primordial human" that makes stone tools and would resemble something like Homo sapiens sapiens.

I think that the biblical reference defines our beginning as humans. Such beginning is of a new species that I would name ***Homo loquens***.

Thus, in my opinion, the **Bible** characterizes the first humans as speaking beings, and in doing so, it **defines humans as a new species that is distinct while mostly disconnected at the cerebral level of any other species inhabiting the Earth**. If that is the intended biblical meaning, it sounds correct to me.

On the other hand, the biblical description, as it seems to me, pictures the "***word***" as an almighty feature, or as the most important human characteristic that helps "***tooling everything else***." But any other "tool" is mentioned there because the language is "a tool generator in itself."

Language occurred due to a forcing geophysical factor that was cosmogenic in essence. I would say that **language contains into its features the provision defining the *cosmogenic genesis***. The universe imposed it like a high-rank alternative that moved us into a higher hierarchy. Here, it remains a question that I would not answer: how long would our language last?

OUR CIVILIZATION SEEMS TO EXPERIENCE A CHRONIC EXISTENTIAL THREAT BECAUSE OF A SEVERE OXYGEN DEFICIENCY

After consulting several hundreds of medical research papers, I came to a significant (for me) scientific conclusion. At some point, I ran into HBOT (hyperbaric oxygen treatment) research that I continued to research into the success of such therapy for various types of diseases. In all, the studies indicate that in each field of experimental medicine to particular kinds of disease or illness, the HBOT could produce in part or totality a remission.

I was surprised by the overall result, and I tried to understand its global significance for our civilization. It was clear to me that hyperbaric oxygen has a role in supplementing our organism with an amount of fresh oxygen.

Why was this supplemental oxygen so necessary and beneficial?

I concluded that, in fact, **our human body experiences a chronic oxygen-deficiency that increased exponentially in recent times**.

Here, I was able to identify **a common cause for such deficiency**: it refers to **our lifestyle**. I can compartment it into four distinct parts

- the indoor lifestyle of the last five generations or so that is the last but the most important outcome for our oxygen deficiency; in developed countries, we have a 90%-93% indoor life, while the developing world is catching the gap at high rates;
- the food intake that first was radically changed as a result of agricultural practices; it changed again in an effective manner due to artificial fertilizers, insecticides, MGOs practices, and novel food processing and packaging.
- the substantial **reduction in human physical activities** as a result of introducing by the Industrial Revolution of various types of machinery; such machinery gradually replaced human and animal energy with mechanical energy provided by fuels;
- social stress produces inflammation, hypoxia, and disease.

All these four branches, characterizing **our modern human lifestyle, incrementally while cumulatively contributed to a significant reduction in the oxygen intake provided to or available to our body systems**.

As a result, **today's human experiences a chronic hypoxia phenomenon that is mostly unnoticed by the research, but it surfaces when the HBOT is applied. Such hypoxia seems to cause the majority of our modern diseases and illnesses**.

In the meantime, medical literature documents the case of **hyperoxia, where an excess of oxygen, resulting from biological processes, is not entirely eliminated by a process of reoxygenation**. So, many toxins and radicals produced during hyperoxia are improperly annihilated by a weak reoxygenation process that also show a lack of sufficient oxygen at the time and place the human body needs that supplemental oxygen. **Such unprocessed waste also contributes to disease**. So, both hypoxia and hyperoxia could equally evolve into disease because human organism rans short of necessary oxygen resources for accomplishing these jobs.

How do we know this reality: it is because the supplemental of hyperbaric oxygen treatment is capable to equally fix both hypoxia and hyperoxia.

I would say all diseases, which can be treated, at least in part, by hyperbaric oxygen, are, in fact, the manifestation of a form of noticed or unnoticed hypoxia and hyperoxia. Both processes show an oxygen deficiency that stimulate disease.

The frightening element is that, in our time, the mentioned phenomenon of **oxygen-deficiency rapidly accelerates**. The most notorious indication of this alarming situation comes from a **recent dramatic increase in human sensitivity to viruses** that currently but immensely troubles us with the **manifestation of the Covid-19 pandemic**.

Such a pandemic imposes human isolation by areal lockdowns, social distancing, and many other severe measures, which all ultimately heart the economies, interrupt international trade and tourism, cause significant stressing among the affected populations. As it appears, the virus of this pandemic is prone to influence, in the long-term, many other systems in the human body.

There is little information yet, but it seems that it may exist the possibility that the human virus, at some point soon, can migrate into animals that will equally devastate domestic and wild animals to unknown and unimaginable proportions.

These threats are real and quite severe, while they may add more harm to our civilization, **exacerbating all other outcomes of the crisis of oxygen-deficiency**.

I like very briefly to discuss the **consequences of oxygen deficiency**:

- the first obvious outcome is the s**preading of disease and/or illness**; the same oxygen-deficiency process produces each occurrence of disease;
- somehow, a novel development that evolved in the last 60-50 years, but it significantly was accelerated in the previous 20 years, it is a process of increased human **susceptibility or a high sensitivity to viruses**;
- another detrimental outcome is a **decrease in neurogenesis** that has very **negative implications that, by now, had diminished cognition and brain plasticity for several consecutive generations**; in the media, this process is popularly called **ideocracy**.

POST SCRIPTUM

After I added five more chapters to this book and three more chapters to the second book (*Augmentation and Illnesses of Civilization*) I pondered on the journey produced by these two books, which are one in my mind, and about the wisdom, I learned while writing them.

I think it is now a good moment to share such wisdom.

As it happens, when you do research, you might have an intuitive thought to follow. But the results may not prove your initial idea, or even they may show that you are going in the wrong direction. For a while, you may find supporting arguments.

All these things happened to me, too.

I discovered for myself that the language is the most precious tool humanity ever invented. Here, I find out that the language is not simply some means for communicating, but also it inflicts planning, control, and socialization.

As a geologist, I was in love with the idea that the associated domain of geophysics may have played a role in shaping the human brain. Amazingly, when I lost all hope that I will be able to demonstrate that, I found the necessary evidence that was around in the scientific literature since 2014. It did not matter how hard I tried the Google searching engine; it never opened that window of information.

So, accidentally I discovered an article about **the role of atmospheric C14 isotope on the functioning of the Dentate Gyrus of the hippocampus**. The abnormally high concentration of C14 isotope was the result of the atmospheric atomic experiments from 1955-1963. Immediately, I connected this to the disturbances generated by the past geomagnetic excursion of our planetary magnetic field. The rest of the story is in the book. But here, I like to add the information provided by an article written by Stephanie Pappas in Live Science (August 29, 2019), while the original study was published in Nature Communications (August 21, 2019). The article comments on new scientific findings in an experiment with quantum superposition: "One can devise this kind of scenario where temporal order of cause and effect are in superposition of being reversed or not reversed" said study co-author Igor Pikovski, a physicist at the Center for Quantum Science and Engineering at Stevens Institute of Technology in New Jersey. "This is something we expect should take place once we have a full theory of quantum gravity."

"What's more, cause and effect could be reversed", Pikovski said. "Imagine two events, A and B, that are causally related". "A and B can influence each other, but in one case A is before B, while in the other case B is before A in a superposition state. That means that both A and B are simultaneously the cause and the effect of each other".

Then, I pondered about the phenomena of thermonuclear genesis and thermonuclear synthesis, and the message of such phenomena being embedded in the supernovae ejection of gamma-rays.

The supernova-e's gamma-rays reach our planet atmosphere and decay in two secondary particles, from which one is C14 isotope that is our concern in connection to cerebral neurogenesis that takes place in the Dentate Gyrus of the hippocampus.

Beyond the effect that extends the overall brain plasticity (including an increased cognition and an improved memory-storage capacity), I pondered if C14 isotope could also bear, in an altered form, the message of the mentioned thermonuclear genesis and thermonuclear synthesis? Could such supernovae message be altered into what we linearly call "genesis"? And then, I questioned a possible scenario, where the (thermonuclear) genesis (as cause) and gamma-rays (as effect) could switch into our brain to a new relationship, like gamma-rays/C14 isotope (cause) and linear genesis (as effect).

However, our brain selects one piece of information from every million of pieces it receives. Hence, our brain has an extremely limited capacity of decoding the rules of the universe, which are those implicated in "*thermonuclear genesis*."

Thus, such selection alters the "mental message" about a type of "*genesis*."

Without knowing the fundamental rules and mechanisms of the universe, humanity has a single possibility: to reinvent the "*genesis-idea*", while factually building it with linear tools and concepts but different than the nature does, and causing an artificiality.

In my books, I explained about a natural phenomenon of augmentation.

Hence, here the attempt to replicate "genesis" is a production process, but has a close relationship to the phenomenon of "augmentation", that is the production's behavior.

In every "revolution" humanity made, from language to domestication, to settlement in villages and cities, to great temples and monuments, to industrial revolution that was followed by subsequent other technical revolutions, to building the culture and art, we empowered ourselves with the will to take "*genesis*" to the next factual level.

It is a great factual achievement that remains fundamentally linear, sequential, weakly reductionist as compared to the surrounding nonlinearity.

I would say, the only technology we ever invented was the language. All other achievements have been corollaries of language: they have been a variety of factual representations derived from the evolution of language.

The language made us think-as-talking, subdued our brain to develop "imaginary-thinking" that turned into the engine that actualizes our "genesis-ideas" that is its production regulated by augmentation.

The "*genesis-idea*" is subliminally inserted into our mind by the bursts of supernovae ejections, and it unconsciously guides us in the process to "*augment*" our "*artificial world*".

CAUSES THAT PRODUCED THE DEVELOPMENT OF LANGUAGE

In part two of this book, I introduced several sensitive arguments, which indicate that the human brain encountered a tremendous change in architecture and functioning around 40,000-30,000 years ago and which changes continued for the next 35,000 years.

In my opinion, our journey, as "human" definition admits, started with Blake Geomagnetic Event placed 126,000-119,000 years ago.

This "event," otherwise called geomagnetic excursion, is a "local **magnetic field reversal** (Class II excursion) with directions flipping quickly to reversed polarity directions, hovering statistically for at least 6,000 years, **and then flipping back to normal polarity"**. (This quote is from Redefinition of Blake Event by Steve P. Lund, Martha Swartz, and Lloyd Keigwin with University of South California, Los Angeles, CA., and Wood Hole Oceanographic Institution, Woods Hole, MA.).

I consider that **this geomagnetic event** had generated particularly potent effects on the hominin brain neurogenesis, **increasing the cognition that resulted in those mental changes that define what we call Homo sapiens**. These changes are accumulated in what is named paternal and maternal Haplogroup A. Homo sapiens was in exclusivity a native of Africa. Climate changes on the African continent forced Homo sapiens to adapt to them by a mutation that generated paternal Haplogroup B.

However, this journey of our new species (Homo sapiens) reached another significant moment during the Greenland-Norwegian Sea Geomagnetic event that is described as a split event because it can be traced in the epoch 70,000-67,000 years ago, but also 57,000-54,000 years ago. The same geomagnetic phenomenon had been defined in NW China as Gaotai Event, evolving in the period 82,800-72,400 years ago, and in the Gulf of Bengal for 86,000-72,000 years ago.

I would say that the **Greenland-Norwegian Sea Event was active, while intermittently, for the entire epoch 86,000-54,000 years ago, that is 30,000 years.**

Why is this geomagnetic event so important? In my view, this was the epoch when the first evidence of art-making appeared in South Africa (dated around 70,000 years ago).

This same epoch represents the beginning of the Out-of-Africa chain of migrations.

I would not discuss here the accidental migration of some hominins in the Middle East that occurred the 120-115,000 years ago during the previous interglacial that was named Eemian.

Some specialists admitted the migrations begun around 70,000 years ago; others have arguments for a cumbersome process occurring approximately 50,000 years ago.

It is to consider that the Mount Toba giant volcanic eruption (around 74,000 years ago) produced a volcanic-winter that changed the climate globally for several years (probably up to 10 years).

As it seems, this catastrophic eruption forced Homo sapiens to retreat inside Africa into the equatorial zones. At the same time, those who already have been moving around the coastal Arabian Peninsula and Indian Peninsula, and where some of them reached Southeast Asia and South China, were significantly affected, and partially eliminated by the impact of the Mt.

Toba eruption. (Archaeological evidence shows very thick ash deposits, which covered the settlements of South India).

During this volcanic event and its following outcomes, the African inhabitants retreated toward the equatorial area and turned isolated there for sometimes. They encountered an equatorial environmental adaptation that generated paternal and maternal haplogroup B.

As we know, all other but numerous haplogroups were created out of Africa.

Mount Toba eruption may explain the gap that was intervening in the migration from 70,000 years ago until 55,000-50,000 years ago.

The mentioned geomagnetic (Greenland-Norwegian Sea Event) event is important because it created new elements of cognition and new skills, which together permitted Homo sapiens to migrate out of Africa. These adaptations were expressed within newly formed haplogroup B. There is to assume that haplogroup B became the motor of migration out of Africa, eventually being combined with the representatives of the old haplogroup A.

* * *

When stepping out of Africa, Homo sapiens needed to adapt to new local conditions, and each such an adaptation generated a new paternal and maternal haplogroup.

Most of the studies of the last decade indicate that the diversity of the local pathogenic environment was the predominant driver of local adaptation of humans to new environments. It was found that microorganisms (microbes but viruses, too) are among the strongest selective pressure encountered by humans.

As Jorge Dominique-Andres and Mihai G. Netea explain in Impact of Historic Migrations and Evolutionary Processes on

Human Immunity (a paper published November 27, 2019, (DOI: https://doi.org/10.1016/j.it.2019.10.001):

"**Pathogens have constantly influenced human migration; therefore, a great number of human genes linked to immune functions and immunity-related disorders have evolved along with humans**."

I would say that most of the human evolution occurred along the human path of Out-of-Africa migration.

I consider that it is reasonable to say that **the effect of the geophysical events became associated with the genetic effects produced by the out-of-Africa migration, which probably together participated in fundamental changes in human brain architecture and resulting functioning**.

Genetic changes in human biological systems directly and indirectly affected the system of command and (neural) control that is the brain, imposing new (adaptive) strategies to all fully developed adaptations.

Here I have to mention that it is a well-documented back-to-Africa migration (sometimes developed later) that significantly contributed to an African genetic remixing that occurred in step with global genetic changes in human biology (imposed by migration).

"Human immune responses have been shaped by the evolutionary pressure exerted by microorganisms and viruses throughout history. Generating a board range of genetic variations and immune functions in different populations favors the adaptation to new environments. It increases the chance of survival of the human species against potential pandemics and pathogens".

"**The combination of the selective pressure caused by infectious diseases with other evolutionary relevant processes**, such as genetic drift, migratory events, bottlenecks, and introgressive hybridization, **contribute to driving the expansion, fixation, or elimination of characteristic immune**

response-related traits in different populations around the world."

Nevertheless, **such a resulting genetic diversity had been instrumental in producing a linguistic differentiation that occurred later during the migration and all further differentiation processes**.

The Out-of-Africa paths of migrants made them confront such novel challenges not encountered inside Africa. The multiple host genes and variants associated with immune functions and disease of the migrants have evolved adaptively at the impact of every such challenge.

The admixture of genes introduced regulatory variants into genomes and caused differences in host immune responsiveness between human populations as they geographically moved around the planet.

Thus, I would say, **human migration out-of-Africa made us so genetically distinct,** when we consider the differentiation that resulted between various human groups, or individuals, or even populations.

Inside Africa, Homo sapiens individuals were confined to one single haplogroup (Y haplogroup A) for something like 50-70,000 years. Only on the brink of the migration, it appeared Y haplogroup B.

Therefore, **each haplogroup appeared only as of the result of multiple migratory events**, and consequently, it would be associated with multiple founder events.

The mentioned studies also showed that **climate only played a relatively minor role**.

The studies detected about 100 genes with a strong correlation between allele frequencies and the novel pathogenic environment. In the meantime, the **genes did not show any similar correlation for diet regimes and climatic conditions**.

I could conclude that human migration out of Africa required strong genetic adaptations to new environments, which represented distinct pathogenic environments. Among human haplogroups, **only the mitochondrial haplogroups show some higher sensitivity to thermal element, like the climate**.

In the meantime, **mitochondrial haplogroups have been assigned a specific role as disease modifiers**. The association of mitochondrial DNA and haplogroups has produced a mixture of mitochondrial functional characteristics and disease itself. The variation in mitochondrial function becomes paramount for the development of the disease, allowing specific association with longevity (in females) and athletic endurance activities (in males).

Human genetics (especially the Y-DNA) were much less influenced by specific climate conditions and the adoption of new diets (climate-adapted diets), where such adaptations made humans better fit in such novel environmental conditions.

* * *

Returning to the epoch that anteceded the language revolution (70,000-30,000 years ago), I have to mention that the **new cognitive capabilities, inflicted by <u>increased neurogenesis</u>, helped develop something like a primitive language** that was instrumental in expressing the first forms of art (cca. 70,000 years ago).

However, this language was very primitive (significantly rudimentary), and it cannot be compared in any way with the language revolution that occurred much later or around 30,000 years ago.

<u>The primitive language expressed until 30,000 years ago had an outcome that was not able to produce a significant distinction between Homo sapiens and the rest of the fellow animals</u>. In many ways, it was similar to the other types of communication (especially the sign languages) occurring between animals or birds. Even the short event of art expression is seen as a onetime only (70,000 years ago) explosion with nothing else following up.

It is clear to me that the **language would not be able to develop unless significant changes would occur in brain anatomy and architecture.**

And such changes were not available until 30,000 years ago.

The question is: were there well-defined causes, which imposed the development of language?

* * *

Comparing with naming visual objects, human ability to name odors is extremely weak; on average, people can call no more than 20-35% of all perceived odors. Humans possess the ability to use language to parse olfactory perception.

Pierre-Marie Lledo and Matt Valley with Pasteur Institute, Paris, France, published in Cold Spring Harbor Laboratory Press, in 2016, an article titled *Adult Olfactory Bulb Neurogenesis*.

The authors explain that "external influences modulate the process of adult neurogenesis at various stages."

They say that "long-term radiolabeling of adult-generated neurons in humans suggested that the **rate of addition to the OB is meager compared to the hippocampus. These data suggest that the canonical process, as defined in rodents**, that adult-born neurons originate in the subventricular germinal zone and migrate rostrally using chain migration to reach the OB circuitry, **may not apply well to humans**. Instead, adult-born neurons in the human OB are likely to be locally generated and few in number, suggesting that **there are significant limits to the magnitude of adult neurogenesis in the human OB**. This identifies **a fundamental difference in the plasticity of the human brain compared to other mammals**".

"Adult regeneration within this structure (OB) underlies some of our capability to learn and associate meaning to our olfactory world."

"Inhibitory interneurons are involved in both the temporal and spatial organization of OB outputs. They transform the spatial dimension of the sensory information reaching the OB into spatiotemporal patterns. **Continuous neurogenesis is, therefore, a mechanism by which the spatiotemporal coding is finely tuned. Adult neurogenesis acts "post hoc" to provide a structural basis for neural plasticity and learning**."

In my opinion, the imposition of an equatorial refuge for African Homo sapiens played an essential role in changes in brain functioning.

The inherited spatiotemporal interpretation of the environment, being generated by the OB, did not work correctly within those new environments, which have significant distinct characteristics. The **spatiotemporal OB comprehension becomes inconsistent in the novel and different backgrounds**. Hence, it needed a parsing that increased comprehension by building a data structure. The parsing process generated a system of symbols associated with OB functions but simultaneously had stimulated the brain segment to develop cognition.

In the beginning, the change or switching was small, but over time, the cognition evolved multifold vs. the OB traditional function.

I would assume that the mentioned incipience occurred 72,000 years ago, and it silently evolved until 42,000 years ago, when another intense radiative burst stimulated it in a fundamentally new manner. This was the beginning (42,000 years ago) of the "language revolution" that developed in association with changes the brain anatomy and functioning, which gradually occurred for the next 35,000 years.

Hence, the **equatorial refuge** forced Homo sapiens groups **to find a new way to improve their comprehension** that was a matter of survival against a relative path of possible extinction. Such a situation required better cognition and awareness of the environment that stimulated an increased sharing capability that attracted a novel group-coordination and activity.

As Ann-Sophie Barwich, with Indiana University in Bloomington, USA, in her paper *A Critique of Olfactory Objects* said:

"Upon encountering olfactory information, the brain essentially has to decide on the grouping, salience, and value of olfactory input by matching this information with learned templates of odor recognition and other cues (cross-modal, verbal), all indicating the context in which a decision about the incoming olfactory information takes place."

"Context appreciation goes beyond direct olfactory determination."

The author explains that the **development of language turned all perception contextual** and **produced a blueprint for the production of "imaginary thinking."**

In the meantime, **it allowed an interpretation of various contexts arising in distinct environments**. The primary initiation of this new mental capacity had occurred during the equatorial refuge era.

It allowed Homo sapiens to comprehend while adapting to various environmental contexts. The general perception became "contextual."

Contextuality was a critical feature developing gradually in our ancestors' brains.

But it would turn into a full-scale approach only in the oncoming era that started 30,000 years ago. **It would become the mechanism that helps create our "imaginary civilization."**

However, **contextuality became the main characteristic of the Late Paleolithic and Mesolithic mind**. **<u>It led to correlative thinking based on the features of contextuality</u>**.

An increased cognition and comprehension implied an increased analysis of information and stimulated the development of a system of group sharing (sign and primitive language) of such knowledge.

The sharing of expertise needed the **creation of a system of symbols, which would intermediate knowledge sharing**. In short, different knowledge became expressed as symbols, which could be communicated and thus, shared.

I would say that the **equatorial refuge led to the development of an increased system of symbols, mostly expressed within a sign-system**, which, many years later, or around 30,000 years ago, would represent the blueprint from where the language would rapidly evolve.

Once the **symbolism was suddenly forced to develop within Homo sapiens mind**, it almost immediately proved to be an extraordinarily useful asset in Homo sapiens adaptation to a variety of other but novel environments.

The **symbolism found expression in the first art achievements**, even when it turned to have a concise life. It is to assume that after its first expression, the symbolism manifested in the art (70,000 years ago) turned into a dormant feature until some 35,000 years ago.

The most significant outcome of symbolism was the need for collective expression and sharing within the group. It increased the socialization in the form of an evolved sign society. It also encouraged the **creation of vocal symbols**, which composed a rudimentary language that spread within Africa, but which was also taken outside this continent by the African migrants.

Such changes ultimately allowed Homo sapiens to migrate out of Africa successfully.

However, at this time (around 50,000 years ago), the Homo sapiens brain was not developed enough to start processing the complexity of a full-scale language.

It was needed another radiative bust to stimulate the neurogenesis in a meaningful manner. But such a needed burst was provided only during the Laschamp Geomagnetic Event that occurred 42,000 years ago.

* * *

During the Greenland-Norwegian Sea lengthy event (almost 30,000 years long), the geophysical disturbance showed a variable intermittence and generated relative lower levels of increased C14 isotope concentration in the Earth atmosphere. The geophysical information shows an increase of C14 isotope atmospheric concentrations at a rate of 20% to 40%.

For example, the Laschamp Geomagnetic Event (around 42,000 years ago) produced a top increase at a rate of 60%-80% in C14 atmospheric concentration and indicated an associated but temporary pole reversal.

As discussed in the previous chapters of part two of this book, the mentioned high concentration of C14 isotope in the terrestrial atmosphere (running from 20% to 80%) inflicted particular changes in Homo sapiens brain.

The entire process, as it has been explained before, stimulated and accelerated the phases of neurogenesis. It **diminished some components of the biological production of nitric oxide that resulted in a hyperbaric-type of blood vascularization of the brain**.

The hyperbaric approach, which causes vascular constriction, is proved to be a central stimulant of neurogenesis.

As the research indicated, **neurogenesis in humans stimulates cognition, while the same process in the rest of animals stimulates only the olfactory glands regeneration**.

There is to assume that somewhere, during hominin evolution, it intervened a cause that switched the role of neurogenesis in hominin brains (from olfactory to cognition). The change was initially tiny, but with time it turned quite large. I **see this cause being generated during the equatorial refuge that occurred 72,000 years ago**.

Each geomagnetic event in the Earth's history produced a weakening of the geomagnetic field and allowed cosmogenic and galactic radiation to penetrate to the ground. While crossing the terrestrial atmosphere, such radiation splits into C14 isotopes and Berillium-10 isotopes.

On each such occasion, because of the penetrating radiation, the atmospheric concentration of C14 isotopes increased from 20% to 80%, depending on the characteristics of each event.

I have to mention that the **Blake Event** (126-119,000 years ago) lasted 6,000 years, and it was an extreme event, causing temporary magnetic pole reversal. It was able to influence the hominin brain significantly, **producing a new species known as Homo sapiens**.

By comparison, the **Greenland-Norwegian Sea event** was significantly less intense, but it intermittently lasted for almost 30,000 years.

I consider that it **influenced the Homo sapiens by producing a primary type of language, and increased the cognition, allowing a successful migration out of Africa**.

However, the Laschamp Geomagnetic Event (42,000 years ago) was followed by a chain of other smaller geomagnetic events for the next 37,000 years.

These **long-lasting intermittent events** (42,000-4,500 years ago) were responsible for producing a chain of anatomical and functional changes in Homo sapiens brains. They **have been so important that, together with the language explosion that followed, created a new species that I called Homo loquens** because of its speaking abilities.

The arguments about this new species are numerous and quite strong.

* * *

Archaeological evidence shows that the process of rounding the human skull is absent before 35,000 years ago. The same evidence points to the fact that the process of rounding was accomplished around 8,000 years ago.

Changes affecting most of the genes resulted in a more spherical shape of the skull that imposed a certain globularity in the form of the brain itself. Such changes in brain morphology led to better neural connectivity. It occurred a bulging of parietal areas and a bulging of the cerebellum.

In the new-born of our species, globularity seems to evolve during a prenatal and early postnatal period. The globularity of the modern brain is directly correlated with both hemispheres of the cerebrum, and both hemispheres of the cerebellum, which together make the entire cranial globe.

The change in shape evolved independently of the brain size. Here, it is known that about 20,000 years ago, the human brain enters a phase of size diminishes.

Globularity is what underlies our species language-readiness. Also, the globularity de-emphasizes the role of frontal lobes as a process that previously has been vastly exaggerated.

Globularity is related to critical neurofunctional consequences. It defines modifications in neural connections.

Most scientists agree that a globular brain gave rise to a language-ready brain. It is known that the frontoparietal-thalamic network emerged only in the context of globularity.

It reduced facial projection that resulted from spheroid shortening, and this has implied changes in oropharynx length, as an adaptation to speech. Previously, it helped humans to develop facial gestures.

Another essential element was how the visual brain switched to a verbal mind. It was a complex process that allowed a general "switching of correlations" and changes the entire functional brain laterality, as described in another chapter.

Globularization has been an adaptation that produced shorter neural circuits, leading to diminishes in energetic consumption.

Other functional neural changes made humans save even more energy by slowing the process of learning.

There no doubt the current human brain is extremely distinct from the epoch of Middle-Late Paleolithic but more critical, humans evolved into a species that is distinct from any mammalian species.

All such fundamental changes occurred only within the neural networks and generated **a new type of brain with new thinking patterns, which both stand to evidentiate for a new species** that is no longer the Homo sapiens species.

This is why I defined it with a proper name that is **Homo loquens** (that translates as **the speakingman**).

ABOUT THE AUTHOR

It investigates the means that creates our civilization, and the means which in response produce illness.

I have an MS degree from the University of Bucharest, Romania, at Geology and Geography Department. I continued to study and I self-trained myself in several other branches of science, like Geophysics, Climatology, Neuroscience, Genetics. Hence, I became a multi-disciplinary scientist.

In my personal life, I worked for almost 40 years as a Geologist Engineer. I am 70 years of age and I continue to work for my own business in the field of Engineering Geology. I immigrated in the US in 1985 from Romania.

I am a member of academia.edu (an online academic organization for academics from around the world) where I uploaded 36 science papers (including the current 2 books in editing and print); I published six books with Authorhouse and three books with URLink Print and Media. Now, this is the 10th book that is the republishing of the mine 9th book to which I added a few more chapters. My papers on academia.edu had over 1,000 views and readers from academics from over 60 countries.

I presented to a conference in Paris, France, two science papers that were the core structure on which this book was made.

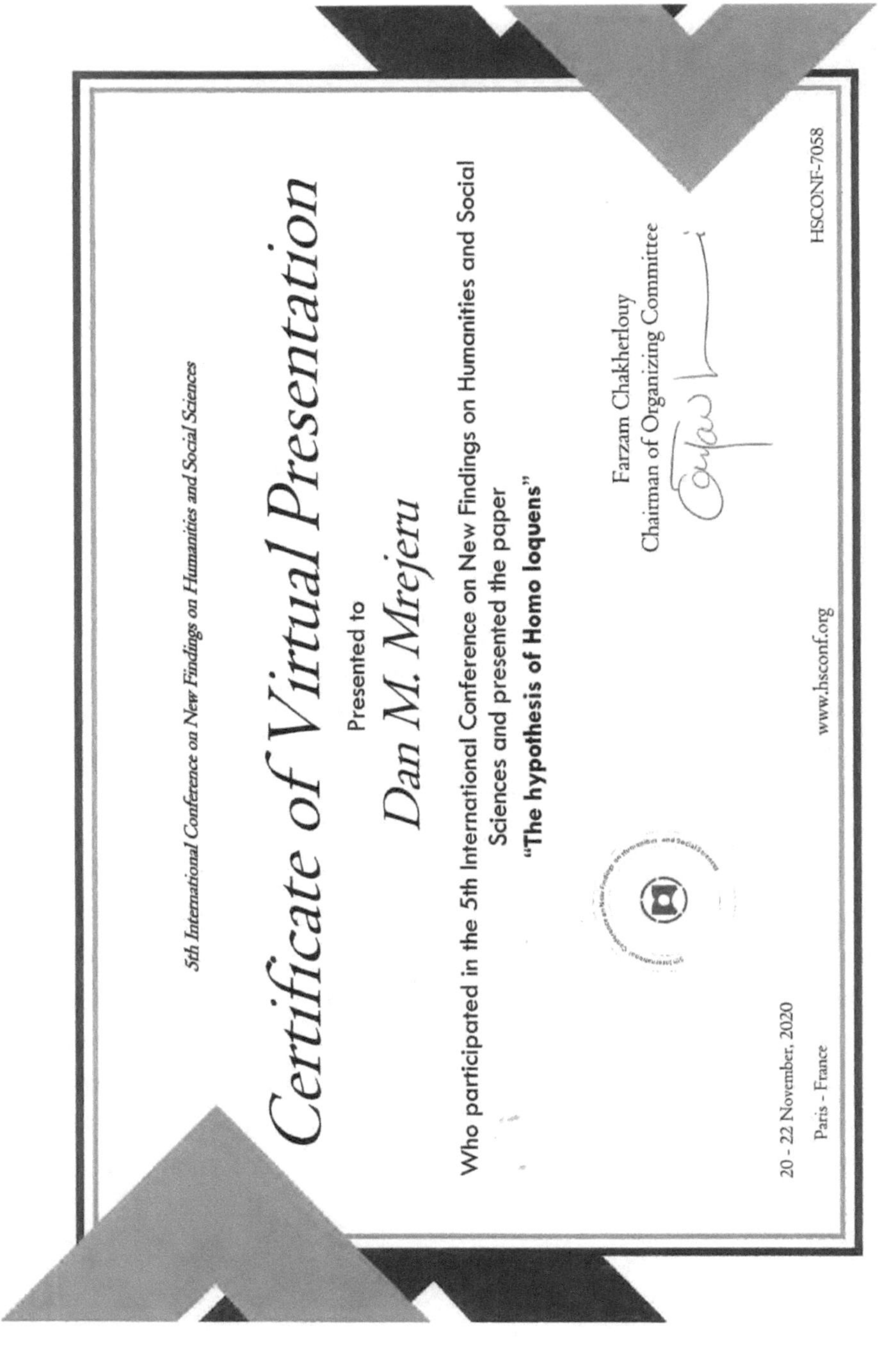

5th International Conference on New Findings on Humanities and Social Sciences

Certificate of Virtual Presentation

Presented to

Dan M. Mrejeru

Who participated in the 5th International Conference on New Findings on Humanities and Social Sciences and presented the paper

"The hypothesis of Homo loquens"

Farzam Chakherlouy
Chairman of Organizing Committee

20 - 22 November, 2020
Paris - France

www.hsconf.org

HSCONF-7058

5th International Conference on New Findings on Humanities and Social Sciences

Certificate of Virtual Presentation

Presented to

Dan M. Mrejeru

Who participated in the 5th International Conference on New Findings on Humanities and Social Sciences and presented the paper

"Our civilization experiences an existential threat caused by a chronic oxygen deficiency"

Farzam Chakherlouy
Chairman of Organizing Committee

20 - 22 November, 2020

Paris - France

www.hsconf.org

HSCONF-7060

9 781953 904218

Printed by Libri Plureos GmbH in Hamburg,
Germany